Fascinated with Science

Fascinated with Science

An Aesthetic Overview of Classical Physics

YUFENG ZHAO

RESOURCE *Publications* · Eugene, Oregon

FASCINATED WITH SCIENCE
An Aesthetic Overview of Classical Physics

Resource Publications
An Imprint of Wipf and Stock Publishers
199 W. 8th Ave., Suite 3
Eugene, OR 97401

www.wipfandstock.com

PAPERBACK ISBN: 979-8-3852-4392-1
HARDCOVER ISBN: 979-8-3852-4393-8
EBOOK ISBN: 979-8-3852-4393-8

VERSION NUMBER 040125

To my wife Sherry, my sons Yihang and Noah

"The heavens declare the glory of God;
the skies proclaim the work of his hands."

—PSALM 19:1, NIV

"Judge the beauties of the heaven and earth,
analyze the essence behind all things,
and investigate the comprehensiveness of the ancients."

—ZHUANGZI, CHAPTER 32

Contents

Illustrations

Preface

AESTHETICS IS OFTEN PERCEIVED as being subjective feelings, termed as "judgement" by Immanuel Kant or "taste" by David Hume. Even when it was defined as the philosophy of arts after Georg Wilhelm Friedrich Hegel, the application of aesthetics has been limited mainly to art works. This situation is in odds with the role aesthetics plays in science, especially physics and mathematics, since classical antiquity. To shed light on both art and science, the theoretical framework of aesthetics that bears the broadest perspective must be constructed within a paradigm rooted deeply to the origin of beauty.

Then why on earth is there such a thing called beauty? Surprisingly, the most influential ancient books seem to say that it is because truth is highly spiritual. More than a decade ago when I was absorbed in close reading the first few chapters of *Genesis*, the text unexpectedly resonated in my heart with the Chinese classic, the *Book of Changes* or *I Ching*. Both classics were venerated for their spiritual value, respectively, in western and eastern worlds. Although these two classics express in totally different cultural context, they contain the same set of universal principles of aesthetics. To many people, *Genesis* 1 is no more than an ad hoc narrative for religion. However, the marvelously elaborated text implies big ideas. In contrast to the literary images in *Genesis* 1, the *Book of Changes* uses symbolic images to link the archetypical images in the natural world to universal principles of reality. These principles are self-evident to artists, therefore, have been applied to the creation of all forms of art and literature for thousands of years.

Based on the above ideas, this book presents a first-principle theory of aesthetics and its application to criticism of the three fundamental branches of classical physics—mechanics, thermodynamics, and electrodynamics. Each branch of theory is treated as an aesthetic story. The development of the story observes the aesthetic principles which hold the key to transparent understanding of the core concepts and fundamental laws of physics.

This book was written initially as an introductory handbook of basic concepts in physical science for Christian liberal art colleges. The author hopes it may also be inspirational to science teachers and graduate students who are interested in philosophy of science.

Salem, Oregon, 2025

1

Introduction

THE RISE OF MODERN science in the western world during the last few centuries deeply changed human thinking, showing the power to predict the intricate structure of the physical universe. The epistemic pilar of modern science is logico-empiricism (or logical positivism) based on mathematical methods and scientific experiment. On the other hand, in sharp contrast to the highly realistic approach of logico-empiricism, the idealistic epistemology of Plato and Pythagoras breathed into modern science the spirit of idealization[1] in the form of thought experiments,[2] which plays a critical role in making discovery in physics. As a result, the heritage of classical aesthetics was proclaimed constantly by the founders of modern physics,[3] whose aesthetic criticism of classical theories nourished their radically creative ideas leading to the scientific revolution.

1. Mizrahi, "Idealizations," 237–52.

2. See Gendler, "Thought Experiments," 1152–63; Elgin, "Fiction," 221–41; Stuart, *Routledge*, 526–44.

3. See Poincaré, *Foundations*, 321–54; Penrose, "Role of Aesthetics," 266–71; Heisenberg, *Frontiers*, 166–83; Dirac, "Excellence," 41–6; Chandrasekhar, *Truth and Beauty*, 59–73.

1

The proclamation of aesthetics by pioneering scientists in history has been revisited in recent years and studied by philosophers with mixed feelings.[4] The philosophers who took the burden to study aesthetics for science were surprised at, if not irritated by, scientists' obsessing with beauty rather than empirical verification, as reflected in the famous assertion by Dirac: "it is more important to have beauty in one's equations than to have them fit to experiment."[5] This assertion was interpreted as "beauty is more important than truth",[6] which is likely based on the presumption that empirical criteria is the highest standard of truth. But, for most practicing physicists, an equation fit to experiment is just an empirical model, which should by no means be called truth itself. On the contrary, an equation with beauty and simplicity is more likely rooted in fundamental laws that represent the highest truth in physics. Just like Chandrasekhar put it, "a theory developed by a scientist, with an exceptionally well-developed aesthetic sensibility, can turn out to be true even if, at the time of its formulation, it appeared not to be so."[7]

The dispute between philosophers and physicists is not a "cultural conflict". It deeply reflects the tension between two opposing views of truth and beauty, namely the classical idealism of Plato and the modern empiricism advocated by David Hume et al.[8] Classical idealism proclaims objective beauty, as represented by the Platonism[9] and the Pythagorean approach.[10] In Humean view, however, beauty is subjective.[11] By putting the secondary

4. See Engler, "Aesthetics," 24–34; Kivy, "Science," 180–95; McAllister, "Beauty," 61–9; Zee, *Symmetry*; Walhout, "Beautiful," 757–76; Breitenbach, "Aesthetics," 83–100; Melo, "Aesthetic Criteria," 96.

5. Dirac, "Evolution," 45–53.

6. Todd, "Unmasking," 61–79.

7. Chandrasekhar, *Truth and Beauty*, 59–73.

8. Melo, "Aesthetic Criteria," 96.

9. See Melo, "Aesthetic Criteria," 96; Field, "Plato," 131–41.

10. See McAllister, *Beauty*, 61–9; Breitenbach, "Aesthetics," 83–100.

11. See Breitenbach, "Aesthetics," 83–100; Melo, "Aesthetic Criteria," 96; Todd, "Unmasking," 61–79.

reality (i.e., fragmented facts or empirical observation) above the primary reality (ideal of physics represented by laws of nature), Humean empiricism rebels against classical idealism and inspired both logical positivism for science and projectivism for aesthetics, therefore ruptured truth and beauty. Logical positivism[12] is a prevailing ideology in contemporary societies of both philosophy and science. Nevertheless, the practice of physics in foundational level seemed to guide leading physicists intuitively to adapt their thinking to the classical idealism,[13] like the case that Einstein was "converted" to Platonism.[14] The fact that aesthetics is appreciated most by the greatest physicists indicates that the deeper scientists dig into the foundational level, the more surprises and excitement of exploration would convince them of the objective beauty of nature.

The tension between empiricism and idealism stretches the current study on the aesthetics of science in a corresponding spectrum. Major theses are recognized in literature standing for the state of the art: the empirical induction of aesthetics by McAllister,[15] the Kantian proposal by Breitenbach,[16] the sensory-dependent formalism by Zangwill,[17] the classical aesthetics combined with realism by Kivy,[18] the theological view,[19] and the classical or Platonic view.[20] Although the models proposed are valuable in a broad field of humanity, it is not sufficiently practical for application to physics. It has been criticized that "the vocabulary of aesthetic appraisal employed in mathematics and science is relatively impoverished

---

12. See Friedman, *Reconsidering*, 1–14; Godfrey-Smith, *Theory and Reality*, 19–38.

13. Heisenberg, *Frontiers*, 166–83.

14. See Melo, "Aesthetic Criteria," 96; Norton, "Nature," 135–70.

15. McAllister, *Beauty*, 61–9.

16. Breitenbach, "Aesthetics," 83–100.

17. Zangwill, *Metaphysics*.

18. Kivy, "Science," 180–95.

19. See Walhout, "The Beautiful," 757–76; Wolterstorff, *Art in Action*, 65–174; Polkinghorne, *Serious Talk*.

20. Melo, "Aesthetic Criteria," 96.

compared with the rich repertoire used in relation to art."[21] Many terms currently used for aesthetic criteria might be called "bounty words that announces and promises some benefit that cannot be controlled or measured".[22] Notice that words such as unity, organicity, coherence, order, symmetry, elegance, simplicity, harmony, charm, etc. are largely the manifests or taste of beauty. Going beyond these superficial descriptors and digging into the concepts deeply related to the origin and basic ways of expression of beauty, ought to be the right direction towards a first-principle aesthetics that should be formulated with rigorousness, unambiguity, and uniformity.

Surprisingly, such a theory is hidden in *Genesis 1* in the Christian Scripture and the oldest and most important Chinese classic, *I Ching* (or *the Book of Changes*). As the cognitive fruits of the Axial Age,[23] the Chinese classical idealism and Platonism have a nearly identical metaphysical foundation. But the formulation and vocabulary are totally different, and *I Ching* may be more aesthetically oriented. Most importantly for the current study, the primary derivations of *I Ching*, based on its first-principle metaphysics centering Tao, can be viewed as fundamental principles of aesthetics. These principles rooted in Tao are essentially self-evident to artists, therefore, have been applied to creation of all forms of art and literature for thousands of years in China.[24] Owing to the spirit of classicism and the contrasted attributes of the entities, the approach is clear and tangible for scholarly treatment. Most remarkably, the beautiful elaboration and rigorous format of the literary narrative in *Genesis 1* cast light on *I Ching* in such a way that the enigmatic statements of *I Ching* become highly transparent as being informed by *Genesis 1*.

This book is arranged in the following chapters. In the next chapter, the basic derivations in *I Ching* are introduced. Five universal principles of reality are extracted out of the primary

21. Todd, "Unmasking," 61–79.

22. Rota, "Phenomenology," 171–82.

23. Jaspers, *Origin*, 8–29.

24. Shih, *Literary*, 8–16.

derivations. These principles are installed as axioms of aesthetics. Theorems can be further derived from the axioms. Within this theoretical framework, a scientific theory can be treated as an aesthetic story. The task of aesthetic criticism is to analyze the aesthetic order in the development of a story, which represents the power of enlightenment for transparent understanding of the concepts, principles, and the structure of the theory. Chapters 3–5 are devoted to criticism of the three fundamental branches of classical physics, namely, classical mechanics, thermodynamics, and electrodynamics. In chapter 6, a general overview of the three branches in a whole picture is provided. Implementing the present theory in development of artificial intelligence (AI) may facilitate more organic creativity of AI systems.

2

A First-Principle Aesthetics in *Genesis 1* and *I Ching*

The Origin of Beauty and the Primary Derivations of I Ching

A COMPELLING DISCUSSION OF the origin of beauty can be recognized in the "happy fish debate" between two ancient Chinese thinkers Zhuangzi and Huizi.

> Zhuangzi and Huizi were strolling along the bridge over the Hao River. Zhuangzi said, "The minnows swim about so freely, following the openings wherever they take them. Such is the happiness of the fish."
>
> Huizi argued, "You are not a fish, so whence do you know the happiness of the fish?"
>
> Zhuangzi said, "You are not I; so whence do you know I do not know the happiness of the fish?"
>
> Huizi said, "I am not you, to be sure, so I do not know what it is to be you. But by the same token, since you are certainly not a fish, my point about your inability to know the happiness of fish stand intact."
>
> Zhuangzi said, "Let's go back to the starting point. You said 'whence do you know the happiness of the fish?' Since your question is premised on your knowing that I

know it, I must know it from right there, up above the
Hao River!" [1]

This parable amusingly yet profoundly reveals the clash between the two views about the origin of beauty. Huizi was a "sub-
jectivist" who believed that beauty is a human feeling. Zhuangzi,
as the leader of Taoism School after its founder Lao Tzu (or Lao
Zi), deemed that Tao is the source of beauty. When Huizi ques-
tioned his opponent, he did not realize that both human language
and the imagery language of nature (e.g., the swimming fish) are
inspired by Tao. According to the Chinese classical idealism, Tao is
the ultimate truth, and beauty is the way Tao expresses itself. Truth
and beauty go side by side. This is reflected in the famous quote of
Zhuangzi: "Judge the beauties of the heaven and earth, analyze the
essence (deep truth) behind all things, and investigate the comprehensiveness of the ancients".[2] Then what is the way that truth is
expressed? The answer is the "change" as defined in I Ching.

The primary derivations in I Ching says, "Therefore, in change
there is the great ultimate. This is what generates the two modes.
The two basic modes generate the four basic images, and the four
basic images generate the eight trigrams".[3] These derivations of I
Ching, shown in Fig. 1, stand for the metaphysical foundation of
Chinese classical idealism. Obviously, these statements of I Ching
are highly enigmatic for most modern readers because of the symbolic images it uses. However, the logic structure of Genesis 1 shed
light on the basic derivations of I Ching, as shown in Fig. 1.

1. Zipory, Zhuangzi, 141–2.
2. Zipory, Zhuangzi, 267.
3. Lynn, Changes, 65–6.

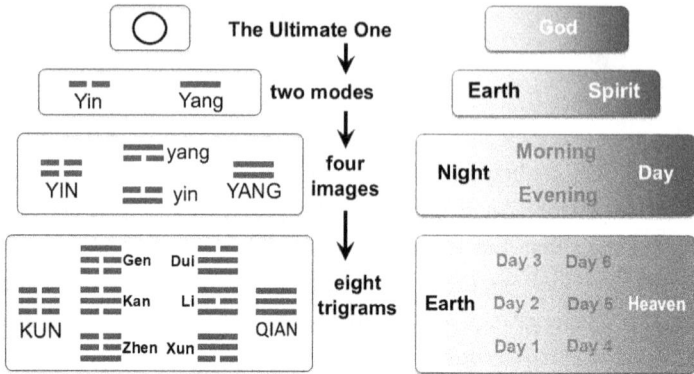

Figure 1. Primary derivation in *I Ching* in comparison with the six-day creation of *Genesis 1*.

The Great Ultimate, namely Tao, is the ultimate truth and the highest unity (One) of the ideal world. Therefore, the Chinese classical worldview is monism, not dualism. Then how is the One "split" into two modes, YIN and YANG, which polarize all the following derivatives? This is because the sharply oriented character of truth automatically opens the space for falsehood. As the ultimate truth, Tao is utterly different from anything else. Tao sustains all things and expresses itself in all things, but in the meantime confronts all things which, apart from Tao, have no essence by themselves. Such a tension between Tao and all concrete objects relying on Tao is the "mystery of mysteries, the gate of all wonders" (*Tao Te Ching* 1:5).[4] Therefore, the origin of beauty is not harmony, but tension that brings contrast/conflict. This leads us to the first axiom below.

Paradoxical Reality and the Fundamental Tension: the Two Basic Modes

I Ching says: "Therefore what is prior to physical form pertains to Tao, and what is subsequent to physical form pertains to concrete

---

4. Mair, *Tao Te Ching*, 45.

objects."[5] The first axiom is about the tension between Tao and physical reality or "concrete objects".

Axiom I: The highest truth Tao (YANG) is not observable or testable, the ultimate essence of the observable physical reality (YIN) is not in itself but is in Tao; Such paradoxical reality creates fundamental tension that drives and governs changes of all concrete objects, through which Tao is constantly expressing itself.

In aesthetic terms, paradoxical reality states that the Virtual (ideal) governs the Real, the Real is less certain than the Virtual. This is evidenced in modern physics. The theory of relativity observes that the absolute power of fundamental laws rules over the relativistic behavior of matter, space, and time. Similarly, in quantum theory, the deterministic nature (certainty) of fundamental laws dictates the probabilistic behavior (uncertainty) of particles. Both theories imply that the invisible laws of physics are higher reality than the visible form of matter.

Paradoxical reality is the source of wonder because it creates the fundamental tension between truth and falsehood. Here "tension" is a technical term of aesthetics and featured by two opposing yet mutually dependent components.[6] In *I Ching*, the fundamental tension is presented as the two basic modes (see Fig. 1). Fundamental tension drives the development of an aesthetic story by polarizing all aesthetic objects into opposing characters and constantly forging conflicts. Contrast and conflicts are eye openers because contrasts are signs that carry meaning. Without contrast, beauty in deeper layers of a story can hardly be appreciated. For example, the indeterministic behavior of quantum particles was not considered as beautiful,[7] as compared to the deterministic picture of classical particles. However, when the indeterministic behavior of particles contrasts with the deterministic power of physical laws, our eyes are opened to see paradoxical reality. This

5. Lynn, *Changes*, 67.

6. Tate, *Man of Letters*, 64–77

7. McAllister, "Is Beauty a Sign?" 174–83.

viewing angle of Tao sees objective beauty. As we will see later, such a basic tension forges many conflicts and excitement in physics and chemistry.

Universal Cyclic Paradigm: the Four Images

Axiom II: The reconciliation of the fundamental tension is represented in a logic loop of the four basic images. The loop must have such a direction YANG → yin → YIN → yang → YANG so that the cycle is initiated by and attributed to YANG representing Tao.

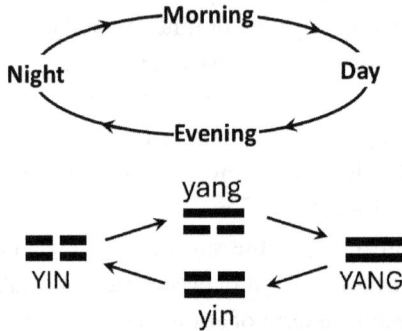

Figure 2. The universal cyclic paradigm in *Genesis 1* and *I Ching*.

Conflicts are created when truth confronts falsehood. But truth does not contradict itself, the conflicts in the aesthetic story must be reconciled in harmony. This is achieved via the universal cyclic paradigm (UCP) of the four images shown in Fig. 2. The four images are highly abstract and symbolized expressions created uniquely in *I Ching*. It is the abstractness that made it a universal principle applicable to all things. The archetypal UCP is seen in the cycles of "Day → evening → night → morning → Day" wrapping up all six-day creations in *Genesis 1*. Most changes in the universe observe the UCP to ensure a changing system is intact. In physics, vibration as the most widely existing form of motion is required by the UCP.

But the UCP concept is far beyond physical processes. *I Ching* shows us how to see the most extraordinary truth through ordinary natural phenomena. Tao has the power of sustainability "because Tao bring renewal day by day that we refer to it here as 'replete virtue.'"[8] The archetypal UCP signifies the deepest UCP of "Tao → generation → the natural world → regeneration → Tao". The conventional explanation of the UCP of the four images was of great ambiguity. Inspired by *Genesis* 1, we proposed here an abstract interpretation: the symbol "YIN" represents falsehood, "yang" can be called testifier, "yin" called justifier, and "YANG" represents truth. For example, Day (YANG) is truth because it is defined by light. Night (YIN) is falsehood because it is where there is no light. Through evening (yin) or fading of light, day justifies (in a negative sense, judges) the night; through morning (yang) when light appears in darkness, night testifies to the day. Similarly, the UCP "truth → justifier → falsehood → testifier → truth" can also be expressed as "essence → informing → substance → manifesting → essence." And this is further concretized for materials science and chemistry, "codes → mechanism → structure → property → codes", which puts the structure-property relationship on a solid basis. Here the "codes" are codified laws of physics in quantum mechanics which arranges all elements in the periodic table. Generally, UCP unifies a scientific theory with four central concepts. Each of the three branches of classical physics, that is, mechanics, thermodynamics, and electrodynamics, has four fundamental laws that are logically arranged in a UCP.

Eight trigrams are further derived from the four images (Fig. 1). QIAN (heaven) and KUN (earth) are called parent trigrams, which give birth to the three pairs of child trigrams. Required by the fundamental tension, QIAN and KUN also represent the two basic modes as the seeds. A whole volume of commentary of the original *I Ching* by Confucius is devoted to explaining the eight trigrams, which is summarized in three central ideas: Yin-Yang is called the way of heaven, Hard-Soft the way of earth,

8. Lynn, *Changes*, 54.

Benevolence-Righteousness the way of man.[9] The third conjugate is a concept limited to humanity, therefore less broad than the first and second ones. In practice, the Chinese tradition often takes Yin-Yang, Void-Fill, and Hard-Soft as three basic norms for artistic creation and aesthetic criticism. The concept of Void-Fill is originally from *Tao Te Ching*, a Taoist treatise based on the original core of *I Ching*. In this article, these three norms are installed as three axioms as the three-dimensional projection of the two basic modes represented by their parent trigrams QIAN and KUN, as shown in Fig. 1. Xun-Zhen corresponds to Yin-Yang, Li-Kan to Void-Fill, and Dui-Gen to Hard-Soft. The attributes of the three (Zhen, Kan, Gen) on the left side are indexed by the position of the Yang line (—), the right three are indexed by the position of the Yin line (- -). The minor party defines the attribute of an image. This is a general requirement in interpreting the trigrams (see Fig. 1) and hexagrams in *I Ching*.

Why does the fundamental tension have these three dimensions? This is because, as we will see in the following sections, they cover three basic issues of changes: Yin-Yang is the driving power and governs temporal order of changes, Void-Fill determines the activity by tailoring free space for change, Hard-Soft regulates the style of changes and forms of matter. Yin-Yang, Void-Fill, and Hard-Soft are shown in their archetypal images, respectively, in Day 1, Day 2, and Day 3 creations in *Genesis 1*.

Yin-Yang: the Way of Heaven

Axiom III: Yin-Yang is about the power of overcoming and governance: Yang represents the power of truth; Yin impedes Yang. The governing power of truth is also the informing power that reveals the truth.

Yin-Yang is the Way of Heaven.[10] Tao is pure Yang (YANG); emptiness is pure Yin (YIN). The universe can be viewed as the

9. Lynn, *Changes*, 120.
10. Lynn, *Changes*, 120.

expression of the former into the latter. Therefore, all things are polarized into Yin and Yang characters. Yin-Yang has a spectrum of meanings that can be projected along two axes, i.e., enlightenment of knowledge (information) and governing power (energy). *First*, Yin-Yang can be falsehood-truth, substance-essence, chaos-order, subordinate-superior, earth-heaven, inactive-active, death-life, etc. *Second*, Yin-Yang prescribes the order of energy distribution, which creates potential to do work, e.g., dark-bright, cold-hot, electric charge polarity, etc. Therefore, Yin-Yang provides driving force. For example, solar radiation (Yang) drives the cycles of material regeneration on the Earth (Yin).

The archetype of Yin-Yang is the contrast of light to darkness as demonstrated by the first day creation in *Genesis 1*. Light symbolizes Tao in many ways. This is the key to understanding aesthetics. Just like that light is both a driving power and a revealing power that facilitates vision, truth is the true light whose governing power is also the informing power that reveals itself by casting light on everything. This is because governance must be achieved in order, and order is furnished with information or knowledge. However, Tao as the ultimate truth and highest revealing power is concealed most deeply according to the paradoxical reality. Therefore, from Yin-Yang, the aesthetic rule of *revealing through concealing* is derived.

Theorem 1: All things manifest themselves in a way of revealing through concealing.

Everything has concealed interior and revealed exterior. In the physical world, the essence is often hidden, and substances are tangible. This aesthetic rule is called revealing through concealing because revealing is achieved through necessary concealing for contrast. For example, optical experiments are often performed in dark rooms so that light images or patterns can be seen clearly. The modern Chinese poet Gu Cheng wrote, "The dark night gives birth to my black eyes, so that I use them to seek the light." Revealing through concealing as a literary rule was expounded by the

Chinese thinker Liu Hsieh in his classical work *The Literary Mind and the Carving of Dragons.*[11]

The rule of revealing through concealing is rooted in the untestable nature of Tao. The fundamental laws stand for Tao in physics because they characterize the fundamental quality of matter, pointing to its ultimate essence. Therefore, fundamental laws are not directly testable. We can only model them through idealization and accept them as fundamental hypothesis. One might argue that Newton's second law is testable, because can be verified by measurement of mass and acceleration. In fact, this is just to test the tangible objects to interpret the law. But the essence of the law is much broader than the observables in its classical interpretation because the origin of mass is not known there. Nor is the nature of acceleration in its equivalence to gravity. Therefore, the purpose of theory of physics is to derive the concrete statement about observables that can be directly tested. As the foundation of science, the descriptive and mathematical forms of fundamental laws are like endless layers of veils over the truth. We peer into its wonder layer by layer but never touch down to the ultimate truth. Such a layered story of scientific discovery is governed by another aesthetic rule that is related with time (history) again through "light".

Light is ordained for visualizing the elapsing time through dynamic contrast to darkness as shown in the UCP of "Day → evening → night → morning → Day". The Chinese poetic term "time" is formed of two characters "light" and "shade". Time is the historical order of changes. All changing things are within time and visible under physical light; but the unchanging laws of physics represent the invisible light, Tao, that governs the visible physical world. The deeper a law is concealed, the greater is its revealing power. Because the ultimate truth is not attainable, the descriptive forms of laws must be updated as our understanding goes deeper layer by layer. The notions developed in the previous layer of the story may be overturned. However, the power and authority of the laws never fails in the scope where they are applicable. This is called canonicity of the fundamental laws.

11. Shih, *Literary*, 214–6.

Theorem 2: Truth reveals itself in layered stories. The self-consistency in each specific layer establishes a canon. Although the notions of the previous layer are superseded or even overturned in the new layer of story, it does not abolish the canonicity of the former layers.

This aesthetic rule is called *novelty in canonicity*, which ensures that the surprise in scientific revolution does not break the consistency between the layers of the story, thus both harmony and tension are fully preserved. In physics, this rule is shown by the correspondence principle.[12]

Void-Fill: the Way of Life (activity)

Axiom IV: Void-Fill is about free space and activity. A space that is totally empty or totally filled cannot facilitate activity. Void represents open space for expression of Tao. Fill is the actual expression via occupancy of space. Void-Fill determines the structure of the story or how an aesthetic story develops.

The second day creation in *Genesis 1* shows the archetypal image of Void-Fill. Just like energy and time are based on Yin-Yang, activity and space are based on Void-Fill. If time represents predestination, then space represents released dimensions of freedom, which is the basis of activity. In physics, momentum (activity) is conjugated with space. Void and Fill are the basic status of space, which prescribes that a physical object occupies a location or state exclusively. Void is not meaningless emptiness; instead, space has format as shown in geometry. Void represents the Virtual, Fill signifies the Real. Paradoxical reality demands that Void preconditions Fill. Void-Fill is the mechanism of activity, functionality, and life. Totally filled space is dead because there is no freedom for change; a completely empty space is also dead because there is no medium to perform function. This idea is articulated in *Tao Te Ching*: "Thirty spokes are joined together in a wheel, but it is the center hole that allows the wheel to function. We mold clay into a

12. Liboff, "Correspondence," 50–5.

pot, but it is the emptiness inside that makes the vessel useful. We fashion wood for a house, but it is the emptiness inside that makes it livable. We work with the substantial, but the emptiness is what we use."[13]

The principle of Void-Fill cast new light on fundamental issues of modern science. Before the 20th century, the whole world believed that atoms were indivisible solid particles. However, Ernest Rutherford discovered that an atom is almost entirely hollow with tiny electrons surrounding a hard nucleus 10^{-15} of the atomic volume. More excitingly, as quantum effect is demonstrated by the electrons (Real), their motion in the real space inside an atom must be formatted by the space of states (Virtual). These eigenstates form electron shells, whose occupancy determines the activity of the atom. An atom or material is active only when the virtual space is open. A noble-gas atom has no chemical reactivity because of its filled electronic shells. For the same reason, an insulator is not conductive because its valance band is fully occupied and closed up by a wide band gap that forbids thermal promotion of electrons to inhabit higher bands in ambient conditions.

Void-Fill opens the space of expression of Tao, therefore determines the possible pathway how an aesthetic story develops, i.e., the structure of the story. Tao expresses itself in the story of "changes". According to *I Ching*, "the changes are generation and regeneration."[14] Reciprocal replication in Yin-Yang is the way of Tao,[15] as illustrated in Fig. 3. This is required by fundamental tension to maintain contrast. Therefore, the primary derivations of *I Ching* (Fig. 1) present the mechanism of generation and regeneration through reciprocal replication of the two modes in increasingly diverse images with multiplication in the number of lines (see Fig. 1), which naturally derives the aesthetic rule of *diversity in unity*.

13. McDonald, *Tao Te Ching*, 37.
14. Lynn, *Changes*, 54.
15. Lynn, *Changes*, 53.

Theorem 3: In the development of an aesthetic story through recipro-cal replication, the diverse objects generated therein are unified in origin, theme, and wholeness of story.

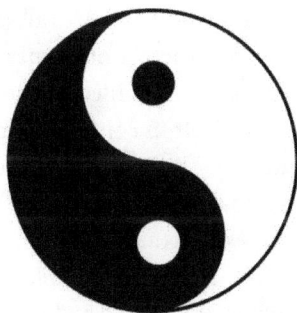

Figure 3. Reciprocal replication of Yin-Yang.

In the derivation of *I Ching*, the images represent the prop-erties of generated objects, and the number of lines denotes the creation of space of expression. Without opening a new space of meaning, no sensible changes of form of matter can be supported. Obviously, generation and regeneration in changes are not a mo-notonous replication but a creative process because opening space of meaning is the key to developing an aesthetic story. In other words, new knowledge of truth must be constantly revealed in changes. Therefore, diversity in unity often goes side by side with another regulation called *strangeness in conformity*, with the crite-rion of strangeness proposed by Francis Bacon, and the conjugated criterion of conformity is formulated by Heisenberg.[16]

Theorem 4: In the development of an aesthetic story, strangeness oc-curs to open more space, but strangeness should abide by and har-monize with a higher conformity.

For example, under the principle of Void-Fill and Yin-Yang, theorems 3 and 4 systematically regulate the properties of chemical elements and the structure of the periodic table. *First*, the

16. Chandrasekhar, *Truth and Beauty*, 70.

Coulomb force between the central massive nucleus with positive charge (Yang) and the wavy electrons with negative charge (Yin) is the governing power that formats the virtual space of the quantum states of electrons. And the derived second-order Yin-Yang measured by the electronegativity polarizes all elements and drives their reactions. *Second*, the structure of the periodic table and the property of the elements are determined by the Void-Fill status of the virtual space (i.e., the electron configurations) under the constraint of charge neutrality. The number of protons or electrons (i.e., atomic numbers), that determines both the strength of the force and the occupancy of virtual space, serves as the index of structural development of the periodic table. *Third*, all elements can be viewed as the replication of the hydrogen atom denoted by "1" representing both unity and activity. Its immediate reciprocal replication is helium (He) denoted by "0" standing for its inactivity due to the filling of the first electronic shell ($1s^2$). Hydrogen has the middle-level electronegativity to be replicated in carbon (C), which results in the nonpolar C-H bond in hydrocarbons playing a crucial role in sustaining all living creatures. Starting from the second row of the periodic table, a new type of reciprocal replication of H generates all the active elements, which splits the electronegativity of H into two extremes, as represented by Li and F. Driven by the gradually strengthened force field due to the increase of atomic number, the first filled shell ($1s^2$) is consolidated with the nucleus into the atomic core and the 2s, 2p states are gradually pulled down, opening more accessible virtual space for electrons. In Li atom, due to the least number of protons and the weakest force field, the atomic core has the largest radius and the single electron in the outermost shell (valence electron) was loosely bound. Therefore, the valence electron in Li is apt to be given away. Namely, Li has the least electron negativity in the second row. Oppositely F atom has the maximum electronegativity because of the strongest force field and smallest atomic core. F atom has an extremely strong capability to get one more electron to fill its outermost shell completely. Between Li and F, electronegativity is diversified in a spectrum from Be to O. And the whole spectrum is divided by C into two regions: to the left is electropositive region and electronegative region is on the right. *Fourth*, strangeness in conformity is observed. For example, the filling of the 1s orbital causes the inactivity of He, sharply contrasts the active Be albeit with the 2s orbital filled. Such a strangeness is harmonized with conformity within Void-Fill: the empty 2p orbitals (Void) in Be pulled down by the atomic core can interact with the 2s orbital (Fill), which provides extra degree of freedom for the two valence electrons of Be. Such a mechanism also facilitates *sp* hybridization. Greater strangeness such as dative bonding and Kubas 2-electron 3-center coordination[17] can also be rationalized within the conformity of interaction between low-lying empty orbitals and fully filled orbitals. Also notice that the positive valence of elements $^+$Li, $^{+2}$Be, and $^{+3}$B is measured by the number of valence electrons (Fill), whereas negative valence of elements $^-$F, $^{-2}$O, and $^{-3}$N are measured by the deficiency (Void) of electrons toward the complete filling. In other words, the minor party of Void-Fill defines the valence.

Hard-Soft: the Way of Earth

Axiom V: Hard-Soft is about forms of matter and style of artistic expression. All the properties that are tangible, observable, or directly testable are related to the form of matter. Forms are sustained by information, which is revealed by imagery signs.

The Chinese philosopher Zhu Xi (1130–1200) said, "When matter is formed of energy, the information must be automatically encoded in its form." [18] Therefore, forms of matter must be sustained by information.

The archetypal image of Hard-Soft is the land profile emerging out of the surface of water, shown by the third day creation in *Genesis 1*. Such an image constantly occurs in Chinese painting and poems as the structuring imagery sign. In *I Ching*, Hard-Soft

---

17. Kubas, "Metal-dihydrogen," 37–68.
18. Zhu, *Commentary*, para 3.

is also called the Way of Earth,[19] because Hard-Soft characterizes physical objects. Form is the tangible way a physical body exists or behaves.

Hard-Soft also has multiple facets of meanings. All those featured by certainty, clarity, rigidity, sharp shape, and being informative can be viewed as Hard; in contrast, the random, mysterious, vague, smooth or round-shaped, and flexible are considered as Soft. Hard-soft governs not only the outer shape of materials, but also the microscopic structure of atoms, all the way down to the fundamental property of matter. For example, the softness of the water is due to the soft hydrogen bonding between water molecules, and the solid land is sustained by the hard covalent, ionic, or metallic bonding. The wave-particle duality of microscopic particles is a manifestation of Hard-Soft in fundamental property of matter. Marvelously, the soft wave motion of electrons confined inside atoms is molded into discrete states which is so hard as to stand against the million-bar pressure in the center of the Earth. This testifies both the hardness of Coulomb force and the rigidity of physical laws that governs the wave behavior of electrons.

Unlike the simplicity of time (Yin-Yang) and space (Void-Fill) as the "framework" of the physical world, forms of matter (Hard-Soft) are substantialized format of space (Virtual) in physical objects (Real). Therefore, Yin-Yang and Void-Fill must manifest in Hard-Soft. Beauty and meaning must be read in forms of expression. That is why all the aesthetic principles and rules introduced above are based on archetypal images of the physical world. The real forms of matter are overwhelmingly diversified with natural irregularity. However, it is natural variability that creates vivid expressions such as demeanor, temperament, charm, fluidity, liveliness, disposition in the soft appearance superposed on the hard characters such as steadiness, firmness, strength, structural elegance, clarity, etc. In Liu Hsieh's theory,[20] the former is symbolically called "Feng" or wind (Soft), the latter called "Ku" or bone (Hard). Therefore, the *theorem of Feng-Ku* is deduced as follows.

19. Lynn, *Changes*, 120.
20. Shih, *Literary*, 162–4.

Theorem 5: The organic conjugation of Hard-Soft is shown by Feng-Ku in forms of matter.

For example, in the fifth day creation in *Genesis 1*, fish has the Feng of three-dimensional freedom shown by its disposition. Birds (Fill in Void) are the reciprocal replication of fishes (Void in Fill). The animals (Hard in Soft) on land created on Day 6 have the Feng of two-dimensional freedom supported by the strong framework of bones and tough muscles, while plants (created on Day 3) as their reciprocal replication have the Feng of waving demeanor out of rigid stem and trunk (Soft in Hard). In Chinese painting, even nonliving objects such as rocks and plants show tension of Feng-Ku in their disposition as shown in Fig. 4.

Figure 4. Disposition: Feng-Ku in Chinese painting.

The criterion of Feng-Ku brings two famous open cases of aesthetics in science to a conclusion. The first is related to the in-deterministic behavior of quantum particles like electrons, which were not appreciated by many scientists like Einstein and Schro-dinger.[21] From the perspective of Hard-Soft, however, the wavy behavior of electrons harmonizes perfectly with the nucleus. This makes an atom matchlessly stylish: the hard nucleus with nearly negligible occupancy of space stretches its powerful invisible force field to mold the soft "matter field" (i.e., electron mist) in spheri-cal harmonic patterns. The second case is regarding the elliptical orbits of planets. Standard circles were considered as perfect orbits for celestial bodies; therefore, the beauty of Kepler's elliptical plan-etary orbits was not appreciated by his contemporary.[22] However, if all the planetary orbits are fixed to perfect circles, the structure of the system would be too rigid, and no strangeness is created as compared to the perfectly spherical force field of the Sun. From this perspective, one would appreciate ellipse as soft disposition (Feng) against the hard spherical force field (Gu).

As a recap for chapter 2, a first-principle aesthetics extracted from *I Ching* and *Genesis 1* is ultimately based on Tao or the Word, whose sharply oriented nature defines the paradoxical reality with the primary being the invisible Tao and all the physical objects being the secondary termed falsehood. Paradoxical reality creates conflict and contrast that serve as eye openers for us to see the truth through its confrontation with falsehood. Such a fundamen-tal tension drives the development of an aesthetic story, which is harmonized by the UCP, "truth → justifier → falsehood → testifier → truth". The UCP shows how truth confronts falsehood through its informing work and how the behavior of the falsehood attests to the truth. The fundamental tension polarized all things into mutu-ally dependent opponents, which are universally characterized by Yin-Yang, Void-Fill, and Hard-Soft. These three principles cover respectively three central issues of the changes in the universe: the driving power and temporal order of changes, the activity

21. McAllister, "Is Beauty a Sign?" 174–83.
22. McAllister, "Is Beauty a Sign?" 174–83.

facilitated by free space for changes, the style of aesthetic objects and form of matter. Under these three principles, five key rules are derived: revealing through concealing, novelty in canonicity under Yin-Yang; diversity in unity, strangeness in conformity under Void-Fill; and Feng-Gu under Hard-Soft. These rules are aesthetic criteria that have been used for a long historical period and only found their firm foundation in the present theory.

3

Classical Mechanics
The Way of Tangible World

NEWTONIAN MECHANICS MARKED THE birth of modern science. The English poet Alexander Pope eulogized the dawn break of this new era by painting a halo of divine revelation on Newton's work: "Nature and Nature's laws lay hid in night. God said, 'Let Newton be'; and all was light."[1] For the poet, Newton had overcome the darkness of the whole universe; for Newton, uncovering all mysteries of mechanics only requires pinning down a few key issues. But to recognize these issues he must be able to see through the two-thousand-year history of human thinking.

The conventional natural philosophy acquired a new life from systematic idealization of natural objects through the work of Galileo (1564–1642), Descartes (1596–1650), and Newton (1642–1727). The new idealism values thought experiment. Furthermore, systematic idealization of physical objects relies on implementation of mathematic methods which is the language of the ideal world. This was powerfully demonstrated in Newton's book *Mathematic Principles of Natural Philosophy*. Idealization made it possible to develop a scientific theory in an axiomatic

1. Cooper, *Physics*, 31.

system, following the example of Euclidean geometry. The great advantage of axiomatic system is that the whole theory is built on a few fundamental hypotheses as first principles to which all the central concepts and their relations are directly tethered. Recognition of these key issues is often a surprising encounter rather than expected results. However, aesthetic criticism as a higher level of thinking often illuminates the process of exploration.

Three Fundamental Issues of Motion

Historians brought up various accounts for the birth of modern science with volumes of fantastic stories. Few of them could circumvent three fundamental issues of motion with their origin traced back to ancient Greek. But why must it be those three? Historians did not tell. However, aesthetically the three fundamental issues of motion are in perfect alignment with Yin-Yang, Void-Fill, and Hard-Soft, as shown in Fig. 5: (i) the framework of the universe based on free space (Void) and evenly elapsing time in which a physical body (Fill) moves; (ii) the deterministic view (Hard) of motion via continuous variation (Soft) of state; and (iii) the cause or driving force (Yang) of motion and its direct effect on a physical body (Yin).

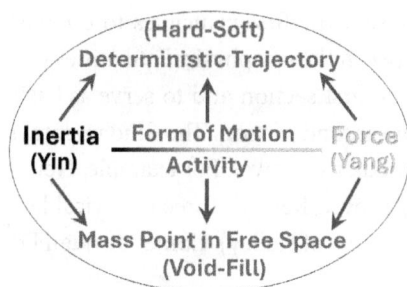

Figure 5. Three fundamental issues of motion and the aesthetic principles of Yin-Yang, Void-Fill, Hard-Soft.

The above three issues turned out to frame Newtonian mechanics in three sections. (i) Idealization of time, space (Void), and moving objects (Fill) opens a clear sky for the development of the new theory. The corresponding ideal models——absolute space and time and mass particle——define the three basic physical quantities: time, length, and mass. Then any other arbitrary physical quantities of motion can be expressed in these three. This is the first section of Newtonian mechanics. Notice that the model of absolute space is simply three-dimensional Euclidean space, and a mass particle is treated as a geometric point. That means the knowledge of the property of space had been made available in Euclidean geometry and the method therein could be directly implemented in mechanics. In addition, analytical representation of Euclidean space in a Cartesian coordinate system was invented by Descartes right before it was needed by Newton. (ii) The second section on the basic form of motion is called kinematics, which is a mathematical description of motion based on continuity of space and time. Instantaneous motion is described rigorously using the concept of differentials. This method was invented by Newton himself, likely having been inspired by the Zeno's paradoxes of ancient Greece. (iii) The third section is called dynamics which deals with the cause of motion. The road to this section had been paved through Galileo's critiques of Aristotle. The core of dynamics is the three laws of motion plus the law of universal gravitation. These fundamental laws laid the foundation of Newtonian mechanics. Also, they represent the first principles to establish the rigorous relationship between the key physical quantities used to characterize motion in the first section and to serve as basis of theoretical derivation in the second section. The fundamental laws endue the theory with a predictive power. For example, Newton's theory not only rigorously proved Kepler's three empirical laws but also predicted the return time of Halley's Comet through Edmond Halley's work.[2]

Although the above theory is built firmly on the laws of motion, a relatively subtle clue to greater wonders is hidden in the level of metaphysics. Newton's conception of absolute space and time assumes that space is uniform, continuously extended,

2. Lancaster-Brown, *Halley*, 78.

independent of matter; time also passes uniformly and continuously, independent of physical processes, as is described in the "General Scholium" of Newton's *Principia*.[3] This view of space and time is a mathematical model allowing straightforward implementation of Euclidean geometry for intuitive description of macroscopic objects at low speed. Newton certainly knew that the status of space and time is inferior to the laws of nature because the laws do not change with space and time. That the laws of nature do not change with relative motion, known as the Galileo-Newtonian principle of relativity, is not a law of physics but an aesthetic principle. It honors the authority of the laws of nature which transcend matter, time, and space. Therefore, the principle of relativity reflects the axiom of paradoxical reality. The aesthetic tension in the metaphysical foundation of mechanics is the impetus for the scientific revolution and eventually inspired the birth of modern physics.

Following the above outline of the three fundamental issues of motion, we will see the story of how Newtonian mechanics develops structurally.

Void-Fill: Matter, Space, and Time

The new idealism of modern science teaches us that the key to finding fundamental laws governing all matter is to reduce the universe to "no matter". This is the way of scientific idealization and how Newtonian physics distinguishes itself from Aristotelian physics. Two ideal models were introduced to depict the basic components of the mechanical universe. One is the model of a mass particle or physical body of matter (Fill); the other is the frame of reference, which is the model of free space (Void). A mass particle is the archetypal object representing essentially every physical body in the universe because mass is a fundamental quality of matter. Ideally, a mass particle is a geometric point with no structure. Only mass is of concern. This idealization circumvents the difficulty of

3. See Cooper, *Physics*, 2; and Newton, *Mathematical Principles*, 6.

dealing with the form of matter that is based on the microscopic structure of substances sustained by fundamental interactions. All physical bodies——from a golf ball to an invisible atom, a planet, or a star——can be treated as a mass particle. A reference frame represents a framework of Euclidean space assumed to be stationary. Newton must have been inspired by Euclid, because his model of the mechanical universe is simply a Euclidean space with the geometrical point carrying mass. According to Euclidean geometry, when the universe with all sorts of shapes is reduced to a few geometrical elements, five self-evidenced axioms emerge to reveal the format of empty space. In Newtonian mechanics, when all matter is simplified to a mass point moving in Euclidean space, the universal laws of motion are revealed, based on which the theory of mechanics is built firmly. Therefore, what is matter itself does not really matter for science. What does matter is the form and behavior of matter that speaks for the laws of nature. The truly marvelous thing is that the ethereal simplification apparently so far off the tangible reality works best for science. This deeply reflects the paradoxical reality.

With models of mass particles and reference frame, motion can be described via the position as the function of time and its first- and second-order derivatives, namely velocity and acceleration. This part of the theory is called kinematics. Most of the quantities involved in motion, except for time, are vectors with both magnitude and direction. Description of motion using average velocity and acceleration is straightforward. However, in terms of universality and predictivity, this empirical approach is highly limited. Rigorous analysis of motion in Euclidean space is appealed for a first-principle theory. It happened that history worked it out in perfect timing.

The mathematical representation of a reference frame is a Cartesian coordinate system. The invention of Cartesian coordinates right before Newton by Pierre de Fermat and René Descartes et al.[4] created a systematic link between Euclidean geometry and mathematical analysis. With Cartesian coordinates, the vectors

4. Kline, Mathematical, 302–21.

used to describe motion can be decomposed into scalar variables and dealt with analytically in each orthogonal direction of the Euclidean space. These orthogonal directions are represented by the axes of the coordinate system. The origin is where all locations are referenced to, and the axes are used to define the independent directions of space and the scale of measurement. The Cartesian coordinates system is a powerful mathematic platform for analyzing the trajectory of motion. For example, both the parabolic trajectory of projectile motion on the surface of the Earth and the general elliptic orbits of planets can be depicted in the Cartesian coordinate system. Once the trajectory of motion is illustrated in space, the continuity of space and trend of motion is clearly seen in a whole picture. However, the trajectory of position is insufficient to describe the state of motion. A meticulous description of its changing rate is required.

Hard-Soft: deterministic doctrine of motion

A systematic theory of mechanics demands a fine description of motion that offers predictive power based on analytical treatment of instantaneous velocity and acceleration. This is the second fundamental issue of motion and proved to be a great challenge. Ancient Greek thinkers clearly had a big debate over this issue, as is reflected in the three famous paradoxes about motion proposed by Zeno of Elea (c. 490–430 BC). For example, the flying arrow paradox states:

> If everything when it occupies an equal space is at rest, and if that which is in locomotion is always occupying such a space at any moment, the flying arrow is therefore motionless at the instant of time and at the next instant of time. But if both instants of time are taken as the same instant or continuous instant of time then it is in motion.[5]

5. Hardie and Gaye, *Physics Book VI*,

Zeno's paradox indicates that the ancient thinkers had clearly realized that motion is not just superficial behavior of physical bodies; instead, it is deeply related to the fundamental property of their existence in space, as seen from Aristotle's commentary.[6] But nobody knew how a body moves closely over an infinitesimal interval of space: is the motion continuously smooth or hopping in unperceivable tiny steps? If it is continuous, is the instantaneous motion certain or not? Many philosophers, including Aristotle,[7] Thomas Aquinas, and Bertrand Russell,[8] commented on Zeno's paradoxes in pure philosophical reasoning. But a practical solution was only given when Newton invented calculus. Aesthetically, this is the issue of changeability vs. rigidity or uncertainty vs. certainty under the principle of Hard-Soft.

Based on Zeno's paradox, Newton was facing two choices to treat motion either as being ideally continuous over space or in the mode of microscopic hopping. The second choice is what mechanics ought to avoid because at Newton's time, microscopic study in experiments was technically prohibited. In fact, Gottfried Leibniz simply rejected it by proclaiming the law of continuity: "nature does not make jumps".[9] We assume that Newton had to choose continuity because that was the only view of motion he could straightforwardly handle. Continuity as the ideal model of motion requires that space and time can be covered by a moving body continuously at any infinitesimal interval. Notice that the two opposing elements tensioning the issue of motion in Zeno's paradox are changeability of the state of motion in a spatial/temporal interval (Soft) vs. the definability or certainty of motion at a specific location/instant (Hard). Numerical experimentation indicates that the average motion over the time interval in the vicinity of an instant is a practical approach to the state of motion at the instant by letting the time interval approach zero. This analysis turns out to be perfectly rigorous and opens a new branch

6. Kline, *Mathematical*, 34–7.
7. Hardie and Gaye, *Physics Book VI*,
8. Cajori, "History," 292–7.
9. Leibniz, *New Essays*, 473–4.

of mathematics, called calculus. Using practical operation to approach the practically inaccessible limit is indeed Galileo's spirit of scientific idealization.

Zeno's paradox reflected the ancient mechanical doctrine of motion: when a particle is at a position it must be rigidly held at the position. Newton replaced it with the dialectically deterministic doctrine: when a moving particle is at a position, it is simultaneously leaving for a new but fully predictable position. Today we know this model is only applicable to the motion of particles with big mass. In quantum mechanics that deals with tiny particles such as electrons, the behavior of the particles becomes fundamentally uncertain. Therefore, the deterministic doctrine must be replaced with the probabilistic doctrine, "a particle *is* and simultaneously *is not* at a specific location because it is probabilistically everywhere." In quantum mechanics, the particle entity (Hard) fuses with the wave property of existence (Soft) into the particle-wave duality within the principle of Hard-Soft. Marvelously, in Zeno's paradox, the phrase "if both instants of time are taken as the same instant" seemingly implies that a body occupying mystically two locations at one instant was indeed envisioned to be an optional ideal. In the probabilistic sense, this is true for a quantum particle. If this is the right reading, most popular reading fails to appreciate the inspirational power of the second half of Zeno's paradox.

Both Cartesian coordinate system and differential method are powerful tools of mathematics that are critical for developing an analytical theory. But the predictive power of mechanics is ultimately held by the laws of physics.

Yin-Yang: Causality and Dynamics of Motion

The third fundamental issue of motion is about the cause and governing principles of motion. This section is called dynamics, based on Newton's laws of motion. The idealistically prescribed laws of physics define the boundary between physics and metaphysics.

Based on phenomenological observation, Aristotle had classified motion into "unnatural" or forced motion (e.g. a moving box on the floor) and "natural" motion (e.g. free falling or rising flames).[10] Aristotle's conclusion on the causality of motion can be summarized as: (i) for unnatural motion, the natural state of a physical body is the stationary state, and an external force is needed to cause motion; and (ii) for the free falling as a natural motion, a heavier body falls faster than a lighter one. There is a basic tension here in causality of motion: if some types of motion need a cause yet others do not, then causality should not be a universal issue as it is intended to be. Aristotle's classification was to reconcile the contradiction by resorting to metaphysical prescription. However, natural and unnatural motion are not the fundamental properties of a physical body but the behavior of motion in different conditions, because the same body can do both natural motion and unnatural motion. Consequently, they should not be installed as universal prescriptions at metaphysics level. In other words, Aristotle introduced too many metaphysical propositions based on phenomenological observations. In contrast, in Newtonian physics, metaphysical propositions are restricted to the foundational level, as implied in the fundamental laws as axioms. All the rest of theory are derived mathematically owing mostly to scientific idealization. This method had been initiated by Galileo who resorted to physical investigation of Aristotle's observations by invoking thought experiment based on idealization, mathematics, and logic reasoning.

Galileo's approach could not eliminate metaphysical assumptions. It only pushed the boundary between physics and metaphysics to a deeper level. To deepen the study, Galileo meticulously distinguished the concept of motion characterized by speed from that of changing the state of motion characterized by acceleration. In his legendary thought experiment, Galileo idealized condition of motion that excludes friction. Then he found that a force does not affect velocity directly; instead, it only affects acceleration directly and consequently affects the average velocity. This foundational clarification is revolutionary because the natural tendency of physical body now is not the stationary state which

10. Cooper, *Physics*, 5.

cannot be well defined; instead, it is the state of motion with a constant velocity. This is crucial for establishing the fundamental concept of inertia and the relativity of motion. With the conception of inertia, all accelerated motions must be forced, therefore, causality is indeed a universal/fundamental issue.

Then, what is the cause of free falling? By introducing the concept of natural acceleration for free falling to supersede Aristotle's concept of natural motion, Galileo could at least treat weight implicitly as a metaphysically prescribed force so that the natural and unnatural motions can be studied on an equal footing. Afterall, intuitively, weight indeed behaves like a force as one can feel it in holding a body. But Galileo and his contemporary might not be willing to call weight a "natural force" which is odd because the word natural means unforced. So, the tension is still there. Through elegant reasoning of free falling, Galileo exposed the tension in a Zeno-style paradox: *if the prescribed force (weight) that causes natural acceleration is supposed to depend only on mass, it turns out that natural acceleration must not depend on mass at all.* Such a statement of tension is called here Galileo's paradox, which can be reached through the following reasoning: if a heavier weight falls faster than lighter weight, the combination of the two is heavier than both, therefore, should fall faster than either one; however, considering the lighter one is lagging behind, it must drag the heavier one to make the combination slower than the heavier; such a contradiction can only be removed by concluding that the heavier weight must fall synchronically with the lighter weight. Galileo's paradox really touched down to the basic tension of mechanics, the tension between inertia and force.

As the impetus of motion, force often appears to be the only central concept in the issue of causality. Therefore, the second law practically overshadows the first law, which is often misperceived as the logical result of the second law at zero force. In Galileo's thought experiments, Newton might have realized tension in Galileo's paradox and recognized that the concept of inertia (Yin) is fundamentally indispensable as the opponent to force (Yang). This explains why he designated the first law of motion to define inertia: *a physical body has the intrinsic property to preserve its inertial state of motion, that is, rest or uniform velocity in a straight line. This intrinsic property is inertia, measured by mass.* The first law as an independent axiom is required by the principle of Yin-Yang.

Because of the inertia, a physical body "refuses" to change its original velocity of motion unless it is forced to. Mass makes a body impeding (Yin) the change of state, therefore, diminishes the effect of force (Yang). Inertia as the inherent property of matter not only links all three laws of motion and the law of universal gravitation but also serves as an important clue to the general theory of relativity. The big wonder in mass is that it not just preserves kinetic energy but also is a locked form of energy, according to the theory of relativity. Energy is a deeper concept of Yang character and will be defined in post-Newtonian mechanics. When energy is released, it turns into force (Yang); when it is locked up as mass, it turns into inertia (Yin) opposing force. Therefore, inertia and force can be viewed as reciprocal replication of mass in the two modes. In fact, such a reciprocal replication is even more straightforward in Newtonian mechanics, because mass is not only the measurement of inertia, but also that of the weight (force). This is exactly what causes the tension behind Galileo's paradox.

According to Galileo's work, the effect of force is to change the state of motion. Newton's second law precisely describes the effect in a mathematical form: *In an inertial frame of reference, the acceleration of a moving object is proportional to the net force exerted on the object and inversely proportional to its mass or inertia.* Because the force of weight acted on any mass particle is proportional to its mass, the natural acceleration of all mass particles would be the same, as shown by Galileo's paradox. In fact, centuries before Galileo, a few philosophers had known that two free-falling balls of different mass set free at the same altitude would hit the ground simultaneously.[11] But the reasoning in Galileo's paradox was much deeper than the observation.

Notice that the direct effect of force is on the acceleration instead of velocity as had been claimed by Aristotle. Mathematically,

11. Cooper, *Physics*, 2.

acceleration is the second derivative of displacement over time. Now the question is, why is it the second derivative, not the first, the third, the fourth, etc.? This question leads us to a deeper layer of mechanics. As we will see, only when the effect of force is on acceleration, Newton's second law can be generalized into the work-energy principle.

When a force is acting on a mass particle, the mass particle reacts to the force with an opposing force. This is Newton's third law: *when one object exerts a force on a second object, the second exerts an equal force in the opposite direction on the first.* This means, action and reaction must take place simultaneously in a pair. For example, when you sit on a chair, you apply a downward force to the chair, which exerts an equal force upward against your body. For a continuum of matter, like a long rope or water in the ocean, action and reaction simultaneously act on any two adjacent parts of the continuum. Through the action and reaction, action of force and motion can propagate in the continuum.

The central concepts in the three laws of motion observes the universal cyclic paradigm: the first law designate mass as inertia (Yin) opposing to force (Yang). The second law defines the action of force over inertia, and the third law testifies force by inertia's reaction to force. Now the only missing link towards a UCP is a first-principle model of force. The discovery of the law of universal gravitation fills the gap in perfect harmony.

Universal Gravitation and the Grand Synthesis

Unlike the three laws of motion that are codes of governance, gravitation is a codified power acting on all matters in whole space. Fortunately, gravitation is so weak that it is literally concealed between earthly bodies except for their weights. Weak as it is, gravitation is universal: *Every object in the universe attracts every other object with a force that is proportional to the products of their mass and inversely proportional to the square of the distance between them. The force acts along the line between the two objects.* This can be expressed in a formula, $F = G (m_1 m_2)/r^2$, where $G = 6.67 \times 10^{-11}$

$N \cdot m^2/kg^2$ is the force constant, m_1 and m_2 are the mass of the two mass particles, r is the distance between them. The gravitational attraction between two 60-kg men is 10^{-6} N, which literally has no effect on their activity. Therefore, gravitation becomes noticeable only when the mass of the object is big enough. It plays a dominant role in shaping the celestial bodies and structuring the cosmos. In the dawning era of modern astronomy facilitated by the invention of telescopes, universal gravitation revealed itself brilliantly over the horizon.

Galileo clearly knew weight-related acceleration in his study. But he named it "natural acceleration" following the context of Aristotle's natural motion, seemingly to circumvent the issue of cause. But Newton should have noticed the contradiction between inertia and Galileo's natural acceleration. Due to inertia, a force between an apple and the Earth is necessary if the apple is falling at an acceleration to the ground from the tree. Probably inspired by Galileo's observation of the rough surface of the moon, which suggests that the heavenly bodies have the same imperfection as those on the Earth, Newton realized that such a force acting on an apple by the Earth should have the same origin as the force that holds the Moon to its orbit surrounding the Earth. Furthermore, the forces acting on the planets by the Sun and the forces between all celestial bodies must have the same origin. This is most likely the initial conception of universal gravitation. The discovery of universal gravitation led to the "grand synthesis" in which the motion of the earthly bodies and that of the celestial bodies are unified in the theoretical framework of Newtonian mechanics. In fact, the solar system is an ideal playground for mechanics because universal gravitation is the only force that dictates the motions with elegant rules shown in Kepler's laws.

Before the discovery of universal gravitation, all forces including weight had only been understood phenomenologically. The importance of the law of universal gravitation goes far beyond gravitation itself in that it establishes the first model of fundamental forces in which the first-principle origin of all complex forces are rooted. For example, weight is essentially gravitation;

and all contact forces such as normal force, elastic force, and friction are ascribed to electromagnetic force. However, gravitational force perceived initially as the "action at distance" was in serious conflict to the widely accepted concept that force must be an intentional action applied through contact in Newton's time. As Newton said of it, "It is utterly inconceivable that inanimate brute matter should, without the mediation of something else which is not material, operate upon and affect other matter without mutual contact".[12] Again, we emphasize that universal gravitation is a fundamental hypothesis which is not directly testable because the ultimate origin of gravitation is not the quest of science but that of metaphysics. Therefore, Newton had to focus on its mathematical interpretation, which implied that the gravitational force is a "codified power". Modern physics shows that gravitation is one of the four fundamental interactions discovered so far. It is these powerful "force fields" that hold the forms of matter and the structure of the whole universe. As quantum field theory becomes the foundation of physics, today we know that not only the forms of matter and their orders of behavior are kept by force fields, but even mass itself is created through the extremely strong force field down to subatomic scale. Therefore, the force-mass UCP is revealed at a deeper level in that the codified power (the Virtual) generates and governs the substance (the Real) under the principle of paradoxical reality.

A Deeper-Layer Story of Mechanics

The conceptualization of "field" is a long historical process after Newton. The earlier stage development was purely mathematical (Virtual) until fields were substantialized through the study of electromagnetism. Since the discovery of universal gravitation as a field of force, free space is no longer truly free. Consequently, physics was directed towards a close investigation of accumulation of the effect of force field over space. For the first time, the "Great

12. Newton, *Papers & Letters*, 302.

Void", which had been the glorious subject of geometry in ancient Greece, became the source of inspiration for physics. The central concepts of physics——mass, motion, force——were gradually superseded by field, energy, and potential. This paradigm shift was transcendently conducted by the principle of paradoxical reality. The development of classical field theory, initiated by the study of universal gravitation, was the main theme of mechanics in the post-Newtonian era, which unveiled a new layer of theory called analytical mechanics. The theory was not only to facilitate the development of thermal physics and electrodynamics but also to pave the road for the paradigm shift towards modern physics.

A key concept initiating the new mechanics is local "action" of a field which observes the Principle of Least Action. Because the force field is a continuum that fills the space, its effect must be described by quantities of motion that are accumulative in space. Therefore, action was defined historically as an integral of a quantity of motion over space or time. The modern definition of action is the integral of energy over time or that of momentum over space. This type of definition formally removes the difficulty to deal with the mysterious "action at a distance". In this context, work is a type of accumulative quantity over a process or pathway in a field. Energy is the state variable that measures the work done by the force field. Although the concepts of work and kinetic energy can be derived from Newton's second law, they only become of central importance in physics with the discovery of conservative force field. This is because force fields, as the codified power of governance, have the capability to do work or store energy. It is the conservative force field that holds the forms of all matters and serves as the ultimate source of energy and information. This idea will be developed more clearly in thermodynamics and electrodynamics, eventually become the central axis of physics.

Work is the accumulative action of force over the moving process, therefore, is defined generally by a path integral of force over displacement. According to Newton's second law, a net force along the same direction of motion accelerates the motion and the force does positive work to the body. We can also say that the positive work of the force causes an increase of speed v of the body with mass m, by which comes forth the conception of the most tangible form of energy, called kinetic energy $KE = mv^2/2$. Therefore, Newton's second law is generalized as the work-energy principle: the net work done to a mass particle equals the increase of its kinetic energy.

Now consider a situation where a driving force does positive work to a body but does not accelerate it. Then there must be an unknown force that is against the driving force doing negative work. Then the positive work done by the driving force is converted to other forms of energy through the negative work of the unknown force. For example, when you lift a body from the ground without accelerating it, the work you have done is converted into a new form of energy called potential energy through the negative work of gravity that is against your lifting. In contrast to the kinetic energy measured by the speed and mass, potential energy is defined by the "potential", or hidden power, of the force field. An interesting property of potential energy is that it can be converted back to kinetic energy, say, when you release the body and let it fall. That means the potential energy is a form of stored or conserved energy. Therefore, the force of gravity that defines potential energy is called a conservative force. The direct consequence of work done by a conservative force is the change of potential energy. When the conservative force is doing positive work like gravity drives free falling, potential energy is converted into kinetic energy; when the conservative force is doing negative work or an external force is doing work against the force field, the potential energy is stored.

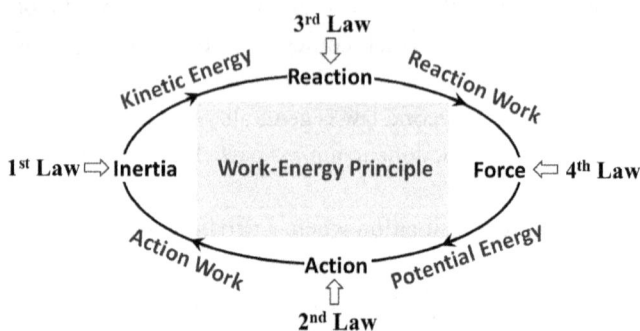

Figure 6. The UCP structure of the four fundamental laws of Newtonian mechanics and its energy perspective.

With universal gravitation as a conservative force being discovered, work and energy serve as the central concepts that harmonize the four laws of mechanics in the UCP (Fig. 6). Powered by potential energy, the force field does work through its action on the mass particle. Then kinetic energy is generated and preserved by the inertia of the mass particle. Kinetic energy is used subsequently to power the particle, and the reaction force does work to the force field. In the whole process, mechanical energy is conserved. This is attested by harmonic oscillation, the most widely existing mode of motion. In principle, the unification of all the four Newton's laws within the work-energy principle is the basis of analytical mechanics. Generally, the work-energy principle deems that work is done through energy transformation. When you push a box, the energy stored in your body is transformed into kinetic energy of the box, which is further transformed into thermal energy through the negative work of friction. Considering all forms of energy overall, the total amount of energy is always conserved. This is the universal law of energy conservation. Energy conservation is based on the fact that all complex forces are ultimately ascribed to the fundamental force fields which do not depend on time.

Just like that work is the integral of force over space, the temporal integral of force is impulse which caused the change of momentum. In parallel to that Newton's second law is generalized into the work-energy principle and energy conservation, Newton's second and third laws lead to conservation of momentum due to the translational symmetry of space. Although energy appeared to be an abstract concept in the early stage of mechanics, it is in fact a fundamental reality of the universe. Einstein's mass-energy relationship reveals that a tremendous amount of energy is needed to create matter, because it costs energy to create the basic order of existence of matter. This again points to the paradoxical reality and fundamental tension: the abstract concepts "order" or "information" are primary. According to thermodynamics, usable energy relies on information.

The profound conceptualization of work and energy can only be fully appreciated when they are tied to fundamental force fields. The highly ordered structure of the universe including the existence of matter is the work of conservative forces. The whole picture will be depicted in chapter 4.

From the anthropic point of view, the codified universal force fields hold all things in order (Hard) but would leave little space for free-will action (Soft). For example, a human body that hosts free will can be easily killed by free-falling in gravitational force field. Wherever the powerful conservative force dominates, free-will activity is prohibited. This is seen from the motion of celestial bodies and the motion of electrons inside an atom. Free-will activity requires that the effect of conservative forces be offset locally by nonconservative forces. Unlike the noncontact force field, nonconservative forces are dissipative due to mechanical contact. Friction is the widely existing nonconservative force in macroscopic environment. Through the negative work done by the friction, the work of the conservative force done to a physical body can be converted to thermal energy or heat. Unlike the potential energy that can be recovered in the form of kinetic energy, heat cannot be converted back into the kinetic energy associated with external motion of the physical body. That means the mechanical energy here is not conserved, instead, it is dissipated into the environment. In fact, the dissipated thermal energy is the kinetic energy related to the hidden random motion of microscopic particles inside the physical body. In contrast to the elegant conservative forces, the dissipative forces are originated from the macroscopic complexity of microscopic mechanisms due to the "nonideal" environment. The dissipative force consumes the work and converts mechanical energy into heat. However, the imperfect (irregularity of) environment creates perfect working conditions that breaks the dominance of force fields and creates a room for free-will activity. We can walk freely on the surface of the Earth and can hold things firmly in our hands, owing to the existence of static friction. This is the beauty of Feng-Ku.

4

Reconcile Yin and Yang
Thermodynamics

THERMODYNAMICS IS THE ONLY fundamental theory of physics
that directly touches the meaning of "meaning" itself. This lies in
the fact that entropy is the only quantity in physics that defines
the degree of disorder. Meaning is encoded in the ordered form of
matter through contrast of signs. The expression of meaning is to
create contrast or tension, which reduces entropy. Besides entropy,
another central concept in thermodynamics, i.e., energy, is mean-
ingful only when there is tension. Without Yin-Yang contrast,
energy cannot even be used to do work. Therefore, the energy
that can do works must be codified under tension. The power that
holds the universe in order is codified through the laws of nature,
particularly, fundamental force fields.

Thermodynamics recognizes the most mysterious wonder
in the universe: the coexistence of two apparently "irreconcilable"
opponents, disorder vs. codified power, which can be viewed as
the tension between death and life. Then, what is the aesthetic origin of such a tension and how is it realized?

Thermodynamic Systems Sustained by Codified Power

Generation and Regeneration: Yang inside Yin

The coexistence of disorder and codified power is aligned with the fundamental tension, signified by Yin-Yang. In the foundational level, Yin points to the empty nature of physicality and Yang represents Tao. By forging tension, Tao expresses itself and gives meaning and essence to all created things. This is called generation. In the meantime, disorder is the natural tendency of the physical systems, which must be sustained by the codified power of Tao through regeneration. Generation and regeneration are realized with Yang running constantly inside Yin. This idea is summarized in *Tao Te Ching* 42:2 as that "all things carry Yin yet embrace Yang; Yin and Yang are reconciled through the interaction of 'breath'(Qi)".[1] Here, Qi or breath represents the flow of energy with information. In *I Ching*, it is shown by the Tai hexagram whose inner trigram is the pure Yang (☰) and outer trigram is the pure Yin (☷).[2] Such a mechanism of generation and regeneration casts new light on the origin of energy and information.

Codified Power: Origin of Energy and Information

Energy and information (Yang), as the "soul" of matter, run through matter (Yin) via the four fundamental force fields. These forces represent codified power that transcends matter. The fundamental forces organize the universe into three major layers (Fig. 7): the elemental or bottom layer in subatomic level ordered by the strong and weak nuclear forces. On top of strong interaction, the electromagnetic force sustains all the chemical elements and governs the chemical reactions that generate millions of compounds.

1. McDonald, *Tao Te Ching*, 80.
2. Lynn, *Changes*, 205–10.

The strong and radical electromagnetic forces are locked up safely inside the atoms and compounds via charge neutrality, while the weakest gravitation reaches out to shape the environment and structure of the Earth, the Solar system, and remote galaxies. The ultimate depth of the "Great Void" is now believed to be rooted in dark matter and dark energy.

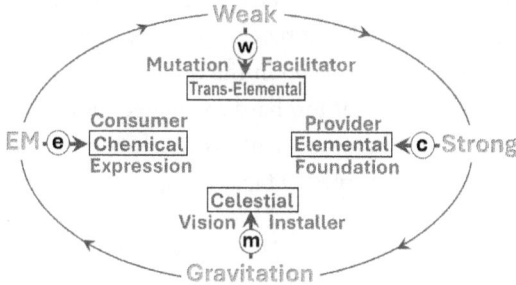

Figure 7. Fundamental force fields, that is, strong and weak nuclear forces, electromagnetic (EM) force, and universal gravitation, as the codified power that holds the universe in layered structures: the elemental, chemical, and celestial layers. Fundamental qualities of matter, i.e., mass (m), color charge (c), weak charge (w), and electric charge (e), serve as the handles of action. These forces dictate the forms of matters and fulfill the purposes of energy and information: the elemental layer is the provider of energy and the elemental unit of information, the chemical layer consumes energy and most actively expresses information in chemical and biological transformation of substance, the celestial layer squeezes the nuclear energy out and installs the energy provider at astronomical distance for safe energy and informative vision.

In such a structure of physical universe, the categorized fundamental qualities of matter such as mass, electric charge, color charge et al. can be considered as matter codes, which is acted on and organized by the fundamental force fields with governing codes such as the symmetry, range of interaction, and strength of forces.

In alignment with the anthropic principle, the middle layer is where the central purpose of the whole structure hangs because this layer is furnished with conditions for free-will activities to be practiced within the governance of laws of nature. A great deal of information is processed and expressed through material regeneration in chemical, biological, and ecological levels, guided internally by the chemical and genetic codes. The material regeneration is powered externally by the solar energy released from the bottom layer. Here the strongest force field with the extremely short range of interaction glues the elementary particles together to form the smallest yet hardest building blocks of matter. Due to the strong force, high energy is locked up in the nuclei. But the strong force makes the elemental layer too hard to change for energy harvest. The weak force, on the other hand, is not for storing energy but to soften the elemental layer by facilitating transformation of subatomic particles through decay. However, the energy harvest from the bottom layer is high enough to destroy all lives, hence must be remotely displaced. It turns out that the energy harvest out of the bottom layer can be carried out through the work of gravitational force in the top layer, then is fed safely to the middle layer over astronomical distance. The gravitational force not only positions the Sun and Earth at the right distance but also squeezes the huge amount of energy out of the elemental layer through gravity-pressed nuclear fusion. The released energy in the form of light radiation warms up the Earth and is stored as chemical energy in green plants, which was subsequently consumed by animals and human beings.

Although the middle layer plays the central role of spokesmanship, the top and bottom layers are also essential for messaging. The elemental layer provides the basic building blocks of matter, therefore, is the foundation of message expression. In addition to the information contained in the forces and hierarchical structures in the elemental (bottom) and celestial (top) layers, the light emitted or reflected from the Sun, remote stars, the planets, and the moons send crucial information to the Earth. Being frequently attracted by the beauty of sunset and sunrise over the horizon, we may not fully appreciate the importance of the global cycles of day and night, and that of four seasons. Such a timing system synchronize all life activities on the Earth and simultaneously create heat flow and temperature fluctuation (or Yin-Yang tension of energy) that powers regeneration of natural resources. Beyond all these, the sublime vision of the sky of lights is a major avenue of aesthetic inspiration, as shown by artistic works.

Based on the above discussion, thermal physics directly studies systems in the chemical layer, which contain ~ 10^{23} microscopic particles. Because each particle is moving in the electromagnetic force field, both kinetic energy and potential energy are stored in a thermal system. The systems evolve and interact with each other through heat transfer, doing work, and exchange of substances. Information measured by entropy is implicitly involved in these processes. Adding heat to a system naturally breaks its order or increases the entropy. Doing work to the system could make it more ordered. The work of conservative force spontaneously organizes a system in order. Gigantic amounts of information were prescribed in the genetic systems of living creatures who can do work to organize their natural environment informed by Tao. The ultimate source of energy and information is Tao.

Aesthetic Features of Thermodynamics

For a long period of history, understanding of thermal effects was limited to the primitive conception of hot/cold states and flowing heat without knowing their nature. This is because thermal effect is the property of thermodynamic systems formed of microscopic particles (molecules or atoms) in random motion within a certain structural network held by the interatomic forces. The molecules are formidable in number and hidden from human sight. No matter how big the thermodynamic system is, the thermodynamic property of the system must be attributed ultimately to the behavior and motion of these microscopic particles.

The first definitive discovery is the mechanical equivalent of heat made by Sir Benjamin Thompson,[3] which indicates that heat is a form of energy rather than a substance. Therefore, heat must

3. Thompson, "XV. New Experiments," 229–328.

be attributed to some type of motion concealed inside physical bodies. This type of kinetic theory had been proposed in its early form by Bernoulli[4] and Herapath,[5] benefited from the laws of motion and the preliminary idea of work and energy conceived in classical mechanics. Here a typical trajectory of scientific exploration is seen in accordance with the principle of paradoxical reality: in mechanics, we discovered the "virtual" (energy and codified power in conservative force field) based on the tangible real objects (macroscopic bodies); the "virtual" then helped us to study the more deeply concealed real objects (microscopic particles) that play a central role in thermal effect so that a new layer of story is disclosed.

Unfortunately, Thompson's idea and the kinetic theory had not been widely accepted until the mechanical equivalent of heat was measured conclusively by James Prescott Joule (1818–1889).[6] This is a crucial step towards a firm foundation of thermal physics. With the nature of heat revealed, the second central concept, temperature, was also found in kinetic theory: in the hot steam that can push an engine turbine, the energy related to the hidden motion of molecules (i.e., the internal kinetic energy) becomes perceivable. Besides raising temperature, heat can otherwise transform a solid into liquid and further into gas. As the phase transition from solid to liquid and to gas, the form of matter becomes less certain (indicating an increase of entropy) and the energy related to the interaction of the molecules (internal potential energy) goes higher, as was envisioned by Rudolf Clausius.[7] Therefore, the key thermodynamic concepts—— internal energy, temperature, heat, and entropy——are defined by the collective behavior of the random motion of the microparticles as described in phase space.

Therefore, filling the phase space with microparticles features thermodynamics: Void-Fill determines the changeability of a thermodynamic system (Fig. 8). Unlike mechanics focusing on

4. Bernoulli, *Hydrodynamica*, 131–42.

5. Herapath, "Physical Properties," 56–60.

6. Joule, "Mechanical Equivalent," 60–82.

7. Clausius, *Mechanical Theory*, 219–21.

the motion of single mass particles in real space, "Fill" in thermodynamics refers to the occupancy of phase space by numerous microscopic particles in thermal equilibrium, and "Void" refers to opening phase space through changing the conditions of equilibrium. Under the principle of Hard-Soft, the form of matter is characterized by the degree of order/disorder of the assembly of microscopic particles. Thermodynamic systems with the natural tendency of slipping towards disordered state (Yin) are sustained and driven to change ultimately by the codified power (Yang).

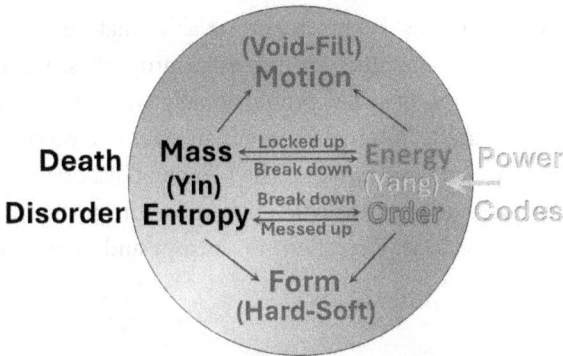

Figure 8. Aesthetic foundation of thermodynamics.

The evolution of thermodynamics systems, often accompanying heat flow and work, is directly driven by thermodynamic potential. Here potential is a concept inherited from classical mechanics, the counterpart of potential energy due to conservative force, measuring the capability to do work. In thermodynamics, potentials are also called free energy functions. *Classical thermodynamics* focuses on the evolution of macroscopic properties derived in differential equations of thermodynamic potential based on the four laws of thermodynamics. *Statistical mechanics* investigates the microscopic origin of thermal properties. *Chemical thermodynamics* integrates the chemical information with the entropy effect to study the direction of chemical reactions.

Thermodynamic Systems: Condition and Characterization

Every arbitrary part of the universe can be viewed as a thermo-dynamic system. All natural processes are thermodynamic ones at the physical level. A thermodynamic system is characterized by variables or its measurable properties. The variables can either describe the state of a system or be related to evolving pathways. For example, temperature T, pressure P, volume V, entropy S, and internal energy U are state variables; Work W and heat flow Q are variables of processes. Properties or variables that do not depend on the quantity of matter are intensive properties. Otherwise, an extensive property scales proportionally with the quantity of matter. For example, internal energy is an extensive variable, temperature is an intensive variable. The following discussion focuses on the central concepts that directly expose the aesthetic facets: temperature and heat (Revealing) vs. entropy and internal energy (Concealing).

Temperature and Zeroth Law of Thermodynamics

Temperature is a directly measurable thermal property because temperature difference determines the direction of heat flow, the key to Yin-Yang reconciliation. However, to define the conception of cold and warm rigorously with the technical term "temperature" calls for a fundamental hypothesis. Like the frequently unper-ceived need for Newton's first law, more so is the law of thermo-dynamics that defines temperature and thermal equilibrium. That is why it was named the "zeroth" law after the first law had been established. Like inertia of Yin character, so is thermal equilibrium as the natural tendency impeding change of a system. The zeroth law of thermodynamics affirms the existence of thermal equilib-rium, i.e., the stationary state of a homogeneous system with all its molecules in random motion.

Thermal equilibrium also applies to multiple systems. When two systems are in equilibrium, the temperatures of the two systems are equal and there is no net flow of heat between the two systems. Also, if two systems are both in thermal equilibrium with the third system, then they are in thermal equilibrium with each other. Based on the zeroth law, a thermometer is designed and used to measure temperature. In the measurement, the system to be measured and the system for standard calibration are both in thermal equilibrium with the system of thermometer.

In fact, all state variables need a well-defined equilibrium so that they can be defined. Generally, for a system in thermodynamic equilibrium, all the state variables can be defined because all the corresponding properties are homogeneous and do not change with time. And there is no net flow of heat or matter.

Although it is empirically tangible that heat spontaneously flows from hot system to cold system, the reason behind this irreversibility is deep. It involves one of the broadest concepts in physics, that is, entropy. Entropy determines the direction of evolution of natural processes. This is the content of the second law of thermodynamics.

Ideal gas, Kinetic Theory, and Entropy: The Mystery of Randomness

A profound and elegant model of thermodynamic system is ideal gas, which treats gas molecules with ideal mass particles interacting with each other only through elastic collision. It is assumed that the mass particle itself has no internal structure.

The ideal gas model is a good approximation for most gases if the temperature is highly above the vaporization point and the average intermolecular distance is sufficiently large, or pressure is low. The ideal-gas system can be described by the ideal-gas law expressed in an equation, $PV = nRT$, where n is the number of moles of gas molecules, $R = 8.314$ J/(mol·K) is the universal gas constant. Absolute temperature scale T in kelvin (K) is defined by this equation. The microscopic version of the ideal gas law is $PV = NkT$ where $k = R/N_A$ is the Boltzmann constant.

Averagely, the random motion of the microscopic particles in a system can be linked to measurable macroscopic properties of the system. This is called kinetic theory, or the broader statistic mechanics when the theory of statistics is implemented. A classic example is Einstein's theory on Brownian motion,[8] which for the first time showed scientific evidence of the existence of atoms and molecules. Einstein estimated that the size of atoms is 10^{-10} m. A much simpler theoretical treatment of ideal gas links the product of pressure (P) and volume (V) to the average kinetic energy ($mv^2/2$) of the molecules through an equation, $PV = (2/3)N(mv^2/2)$. This equation echoes to the ideal gas law given that $mv^2/2 = (3/2)\,kT$. Therefore, the average translational kinetic energy of the molecules in an ideal gas is directly proportional to the absolute temperature of the ideal gas. This is true not only for ideal gas but also for solids and liquids. You touch a hot body and feel burning because the molecules inside the body impact your skin harshly and overheat the cells in your skin.

The above equations indicate that both temperature and pressure (precisely, the product of pressure and volume) of ideal gas can be directly measured by the average kinetic energy of the molecules. The negative product of pressure and volume defines a potential function called mechanical work potential, which measures the capability to do mechanical work to the environment, as expressed in a differential equation, $dW = -\,PdV$. This is how a steam engine does work through pushing the piston. Here P is called a generalized force and dV generalized displacement. P and V are conjugate variables because the product of the two defines a potential function. Now the ideal gas law clearly shows the mechanical equivalent of heat $PdV = nRdT$ at constant volume P.

Similarly, temperature and entropy form another pair of conjugate variables. Here temperature T is the generalized force and entropy S is the generalized displacement. The product TS defines another potential function that can be called heat potential or the capability to store or release heat as shown in the following differential equation, $dQ = TdS$.

8. Einstein, *Investigations*, 19–60.

Entropy measures the dispersal of molecules in phase space. Larger entropy means that the particles are dispersed more "homogeneously". When the entropy turns smaller, the molecules are segregated in some regions of phase space and other regions are sparsely occupied. In accordance with the fact that contrast represents order and information, entropy measures the degree of disorder or loss of information. To understand the effect of entropy, we pick the ideal gas as an example. Due to the entropy effect, maximum dispersal of ideal gas molecules in phase space is always reached in thermal equilibrium with respect to a certain constraint (e.g., fixed volume, pressure, temperature). Then the statistical distribution of state variables can be precisely given. For example, the speed of molecules in an ideal gas system observes the Maxwell distribution with respect to a certain temperature and volume. When heat is added to the ideal gas, the kinetic energy of the molecules must increase. Therefore, the temperature increases, say, from T_1 to T_2. Notice that now the distribution of the speed of molecules is much broader at T_2, according to Maxwell speed distribution (see Fig. 9). This means the molecules are dispersed more "thoroughly" in the phase space. That is, the entropy is larger, hence the system has a higher degree of disorder. Because the temperature and entropy of the ideal gas is higher, the system has a higher heat potential. Also, because the pressure is higher, the system has a higher work potential. In fact, there is a fundamental potential function that measures potential of both heat release and doing work. That is exactly internal energy, U.

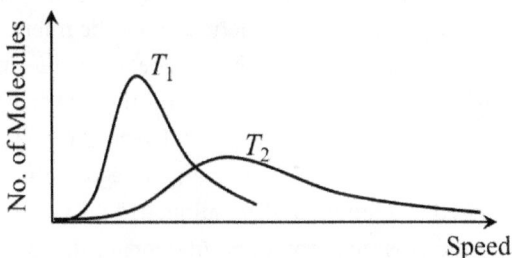

Figure 9. Maxwell's distribution of speed of molecules of ideal gas at two different temperatures, $T_1 < T_2$.

Internal Energy and External Forms of Matter

Because ideal gas has no intermolecular force, the internal energy of ideal gas is the total kinetic energy of molecules measured by absolute temperature T: $U = (3/2) NkT$. Ideal gas is a formless substance except for its mass-point like "molecules" and the external constraint of volume V. Consequently, randomness and dispersal dominate the behavior of gas. This mechanism creates a permeative filling of the space surrounding the Earth with air, which is needed by all living organisms with breath. In sharp contrast, the highly ordered world is full of substances in diverse forms. In fact, every real substance must have external and internal forms, which reveal the ruling force to hold the form against the dispersing effect of entropy. That ruling force is codified power, as shown in Fig. 7. Electromagnetic force holds the structure of atoms and the bonding network that binds the atoms in bulk chemical substances. The mechanism of chemical bonding and the microscopic form of matter are addressed in quantum chemistry. Thermodynamics only discusses the external forms (gas, liquid, and solid states) of substances.

With the intermolecular interaction considered, the internal energy must include internal potential energy. Strictly speaking, internal energy can include any kinds of energy contained in a substance, including chemical energy and nuclear energy. However, in most situations, not all forms of energy are relevant to the thermal process of concern because some of the energy (e.g., nuclear energy) cannot be activated in normal temperature range. From the perspective of kinetic theory, the internal energy is ultimately split into internal kinetic energy of the molecules measured by the absolute temperature and the internal potential energy due to the conservative forces between the molecules. In classical thermodynamics, internal energy as a potential function can be expressed in a differential equation, $dU = TdS - PdV + \mu dN$, where μ is the chemical potential, measuring the capability of chemical reactions driven by the intermolecular forces. TdS and $- PdV$ measure, respectively, the capability to release heat and to do work to the environment. According to kinetic theory, and are the measurements of the average kinetic energy and momentum of the molecules. Therefore, the potential $TdS - PdV$ represents the random motion of molecules and the tendency towards disorder. In contrast, the Gibbs free energy μN represents the codified power or the electromagnetic force that creates both the external forms (gas, liquid, and solid states) and internal forms (chemical bonding) of the substance by organizing its molecules or atoms in order. When the random motion overcomes the codified power, the substance becomes more disordered, and the phase transition goes from solid to liquid or liquid to gas. Otherwise, when the work of electromagnetic force dominates, phase transition goes the opposite way, and the substance is more ordered.

Phase transition is the change of the macroscopic form of matter, observing the principle of Hard-Soft and the principle of Void-Fill driven by the Yin-Yang (see Fig. 8). Thermodynamics explicitly deals with the macroscopic forms of matter which reflect information. Information has multiple aesthetic facets. First, information, as it is encoded in fundamental laws, is the rule of governance, which automatically creates tension or contrast. Tension, in thermodynamics, is the potential as useable energy and capability to do work. Therefore, information has the Yang character, opposing the homogeneity of Yin character. Second, information is expressed in recognizable symbols or codes through contrast (Reveal), opposing the unrecognizable randomness (Conceal). On the other hand, information represents certainty (Hard), opposing uncertainty or disorder (Soft). In cyclic heating and cooling, materials are regenerated in cyclic disassembling and reassembling, so that natural resources can be renewed through the work of codified power.

Measuring Heat: Calorimetry

A straightforward approach to characterization of thermal effects is to measure temperature and heat. This is because substances have a decent temperature-heat property that coordinately calibrates the two quantities, observing the aesthetic rule of revealing through concealing. The temperature-heat property of a substance features its heating process with two characteristic stages: (i) when there is no phase transition, the temperature of the substance increases linearly with the heat added; (ii) during phase transition with adding of heat, the temperature is pinned to a fixed point. Therefore, heat is shown and calibrated by a change of temperature in stage (i), whereas in process (ii) heat is hidden for calibration of temperature itself. The hidden heat during phase transition is called latent heat.

In stage (i), when a substance is heated, its temperature increases accordingly because the heat is converted to the kinetic energy of the molecules. The heat needed to warm up a unit mass of substance by one degree is called the specific heat. For example, the big heat capacity of water is reflected in its large specific heat of 4.186 kJ/(kg·°C), which means warming up 1 kg water by 1°C takes 4.186 kJ heat. The more heat added to water, the hotter the water is because the water molecules move faster. The warming process does not stop until the temperature reaches the boiling point, when stage (ii) starts. The temperature won't increase even if heat is being added. This is because the added heat is converted into potential energy due to the interaction between the molecules. As the potential energy increases, the intermolecular distance is larger. So, the water is boiling, and liquid water transforms into steam. Besides vaporization (a liquid transforms into gas), melting (a solid transforms into liquid) is also a phase transition. During the process of phase transition, the temperature is fixed at the boiling or melting point, and the two phases coexist in thermal equilibrium. Phase transition is reversible, involving the same amount of latent heat per unit mass of substance.

Thermodynamic Processes: Law and Order

Like in Newtonian mechanics, once the central concepts are identified, the fundamental laws are automatically revealed. This is because the central concepts are defined or related to each other through the laws. For example, temperature is defined in the zeroth law of thermodynamics through thermal equilibrium, internal energy is defined in the first law, entropy is the center of second law, and the kinetic theory of absolute temperature is tethered to the third law. These laws govern the interaction and changes of thermodynamic systems as conditioned by heat transfer and energy exchange.

The "Breath": Energy Exchange and Heat Transfer

A thermodynamic system is a scientifically defined object with substance confined in a boundary dividing the system from the environment. Conventionally, based on the property of the boundary, thermodynamic systems are categorized as: (i) an *open system* that can exchange both energy and material with the environment; (ii) a *closed system* that can exchange energy but not material with the environment; (iii) an *isolated system* that can exchange neither material nor energy with the environment. Normally, all objects in the natural environment are open systems. A covered or sealed substance inside a container can be considered as a closed system. Clearly, the above categorization only considers exchange of materials and energy. This may not be complete because the system may receive information. For example, the day-night alternation encodes a message of timing, which synchronizes life activities on the Earth. In fact, Yin-Yang reconciliation always involves information. As we have stressed, living things can be directly inspired or informed by Tao in a way much deeper and richer than a physical system being governed within the laws of physics. This could be a new field of research for living systems.

A system exchanges energy with its environment through heat transfer (Soft) or mechanical work (Hard). Mechanical work is done through changing volume: $dW = - PdV$. Although the work done by heat engines wastefully creates exhausted heat, the life activities of plants and organisms in the forest work together to make the environment healthier because information and order are being put to the soil to maintain the organic composition and structure of soil with balanced pores, carbon fibers, air, water, minerals, microorganisms, etc.[9]

Heat flow is the "breath" for Yin-Yang reconciliation (*Tao Te Ching* 42:2)[10] that sustains all living things through a well-tuned thermal environment. Heat transfer can be realized in three means that cover all situations and across whole space: (i) *Convection*: internal exchange between regions of a fluid system through circulation of the fluid (Soft); (ii) *Conduction*: through the solid interface (Hard) between two systems; (iii) *Radiation*: Across a distance in free space or vacuum. Notice that convection and conduction take place in space filled with matter (Fill), but radiation goes across an empty space (Void).

Leave Sisyphus' Curse for Machines: The First Law and Thermal Engines

A normal thermal system is constantly changing through interaction with the environment. The change can be predicted and controlled purposefully. For example, by predicting the evolving trend of the atmosphere system, we can collect climate information and forecast weather. By designing and controlling the system of burning fuel in a combustion engine, we can use the engine to do work. The development of thermodynamics had been accompanied by the invention and application of heat engines, which initiated the first industrial revolution. Prediction and control of thermal systems are based on the laws of thermodynamics.

The first law of thermodynamics is the law of energy conservation, or the generalized work-energy principle. It states that *the increase of the internal energy of a system equals the heat transferred to the system subtracting the work done by the system to the environment*. For example, in an adiabatic process, there is no heat flowing to or from a system, then the increase of internal energy equals the work done to the system; or the decrease of internal energy equals

9. Dexter, "Advance," 199–238.
10. McDonald, *Tao Te Ching*, 80.

the work done by the system to the environment. In an isometric (or isovolumetric) process, the volume of the system is not changing, so no work is done by the system, the increase of internal energy equals to the heat added to the system. In a heat engine, the internal energy of the system must be recovered to its original level in each operation cycle, therefore, the net heat added to the system can be used to do work. In practice, a heat engine needs two heat reservoirs. One is at a higher temperature; the other is at a lower temperature. Heat flows from the high-temperature reservoir to the operation system; part of the heat is used to do work, and the rest is exhausted to the low-temperature reservoir. No heat engine can possibly convert a given amount of heat completely into work without exhausting heat. This is also required by the second law of thermodynamics. From an aesthetic point of view, the two reservoirs working together in the operation cycle of a heat engine is required by the axiom of Yin-Yang. The high-temperature reservoir represents Yang, the low-temperature one is Yin, and tension between Yin and Yang provides the driving force in the UCP of the heat engine.

One Cannot Step into a River Twice: The Second Law

The second law of thermodynamics governs the direction of thermodynamic processes. We have mentioned the second law twice, accompanying respectively zeroth law and the first law. In alignment with the zeroth law is the *Clausius statement of the second law: no device is possible whose sole effect is to transfer heat from a low-temperature system to a high-temperature system.* Going with the first law is the *Kelvin-Plank statement of the second law: no device is possible whose sole effect is to transfer a given amount of heat completely into work.* In fact, the general statement of the second law is called the *law of entropy increase: Natural processes tend to move toward a state of greater disorder (with less information).* In other words, information cannot be created spontaneously in nature. This statement also prescribes the direction of time: aging is unavoidable in the natural world, and everything is decaying. The ancient thinkers had philosophical statements for the second law. Heraclitus had a quote: "No man ever steps into the same river twice".[11] And Confucius said, standing by the river: "How it flows irreversibly on like this, never ceasing day and night".[12]

The first two statements can be understood from the viewpoint of the third statement of the second law. If the Clausius statement is not true, that is, heat can be transferred from the low-temperature system to a high-temperature system without causing other effects, then sharper contrast is created spontaneously. That means net entropy is decreased spontaneously. To understand Kevin-Planck statement, one should know that work can be used to create order in a system. Therefore, if a certain amount of heat is consumed to do work without exhausting heat, then the net entropy will also be decreased because both consumption of heat and doing work cause a decrease in entropy.

Ultimate Restless: The Third Law

The Third Law of Thermodynamics states that absolute zero kelvin is unattainable. Because temperature measures the average kinetic energy of the molecules in the system, absolute zero kelvin means all molecules are rest.

The third Law testifies the power of conservative force field that holds the molecules in the system. This is because the virial theorem shows that the kinetic energy of the particles in the system is determined by their potential energy. From the perspective of quantum mechanics, zero-point energy is the minimum kinetic energy of the particles that is also determined by the conservative force field. The third Law has an equivalent statement: *no heat engine can achieve an efficiency of 100%.* Such an engine would require an exhausting heat reservoir to be at zero kelvin.

11. Barnes, *Presocratic Philosophers*, 43–62.
12. Chin, *Analects*, Book 9.17.

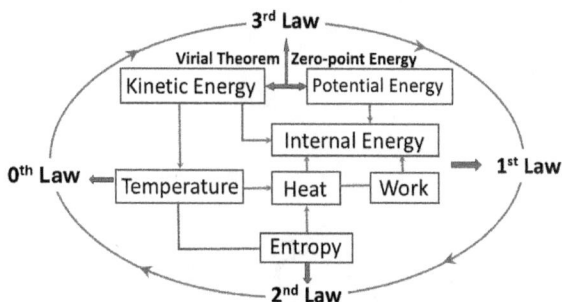

Figure 10. The theoretical structure of thermodynamics: the four fundamental laws and the key concepts are unified by the UCP.

The four fundamental laws of thermodynamics observe the UCP. Thermal equilibrium is state of the maximum degree of disorder, representing "heat death". Therefore, the zeroth law has the YIN character. The first law defines internal energy, which is originated from the codified power of YANG character. Behind entropy is the second law, which is the justifier (yin) that prescribes the natural tendency of a system going from YANG (order) to YIN (disorder). Finally, the third law (yang) testifies to the codified power (YANG) based on absolute temperature defined by thermal equilibrium (YIN). Figure 10 shows the universal cyclic paradigm for the four laws of thermodynamics that unify all key concepts and the structure of the theory.

5

Electrodynamics
Unveil Truth Layer by Layer

The Art of Revealing through Concealing

ELECTRIC CHARGES ARE NOT always readily observed on a macroscopic scale, because they were concealed through charge neutrality at atomic level. However, the power of electricity is revealed overwhelmingly from the sky in lightning bolts accompanied by thunder. Although concealed, electric charges as the fundamental quality of matter often show themselves in trace amount via triboelectric effect, which break charge neutrality by charge separation through surface friction. For example, a finger touching a dry surface may feel electric sparks. Triboelectricity had been studied since Axial Age[1] and was developed in recent centuries to generate massive electric charges which can be stored and discharged abruptly to create "artificial lightning" and helped to unveil the fearful flashing thunder in the sky.[2] The tension between revealing and concealing of electric charges created such mysticity that the word "electrician" once upon a time was almost the synonyms of

1. Benjamin, *History*, 15–20.
2. Joseph, *Lightning*, 35–40.

specialized "magician", referring to those who performed spectacle of electricity to thrill the authorities or the public.[3]

But the application of triboelectricity is restricted by massive charge generation, durable storage in insulated "ports" of charges, and steady delivery of the electric power. Perhaps no one had dreamed of electricity without breaking charge neutrality, which turned out to be the best solution history provided through the invention of batteries based on the new concept called electrochemistry.[4] In fact, both triboelectricity and electrochemistry are based on the electron affinity of materials. Unlike triboelectricity that charge an effective "capacitor" (or ports of opposite charges), batteries use the difference of electron affinity as an electromotive force (EMF) to generate steady electric current in a closed loop of circuit via electrochemical reactions.

Besides the promising practical application of batteries, this great invention had a more profound impact on science. First, the electricity driven by a circuital electric field rather than charge separation promoted the paradigm shift from charge to electric field as the central concept. Second, the creation of steady current paved the road to synthesis of electromagnetism —— the unification of electricity and magnetism. The similarity between electric charges and natural magnets had been noticed in classical antiquity.[5] However, the bridge between the two had never been built before electric current was steadily generated. This is because static charges do not interact with a magnet. Upon the invention of batteries, a great discovery was made by Hans Christian Oersted (1777–1851) that electric current induces magnetic field and interacts with a magnet.[6] Therefore, magnetism was found to be the effect of electric current (Ampere's circuital law),[7] rather than a fundamental quality of matter. It is emphasized that charge neutrality in steady current is critical for revealing magnetic field

3. Joseph, *Lightning*, 22–31.
4. Brett and Brett, *Electrochemistry*, 13–38.
5. Benjamin, *History*, 15–20.
6. Oersted, "Experiments," 273–6
7. Williams, *Michael Faraday*, 143–50.

through concealing electric field. For example, when two steady flows of electric current interact with each other, the interaction between them cannot be interpreted as the interaction between net electric charges because the latter are obscured by charge neutrality. Instead, it must be interpreted as the interaction through magnetic field induced by the current. Consequently, as the electric origin of magnetism weighed in, the central paradigm shifted further towards field theory.

The synthesis of electricity and magnetism into electromagnetism broadened and deepened the study both scientifically and technologically. If we view steady current as charge separation in the reciprocal space,[8] then electric fields and magnetic fields are the result of charge separation in, respectively, real (x) space and reciprocal (k) space. This is to say, from the aesthetical viewpoint, magnetism and electricity are reciprocal replications. Because electric fields cause charge separation in real space, we should ask the question: would magnetic fields be able to cause current or charge separation in the reciprocal space? The answer turns out to be yes (Conformity), but through a totally different story (Strangeness) from the real-space charge separation caused straightforwardly by an electric field. This is because a magnetic field is not a driving force for electric charges. To motivate electric charges, it must first induce an electric field. Michael Faraday (1796–1867) discovered that, (i) a magnetic field may induce an curling electric field which drives electric current in a closed loop of circuit, similar to the electromotive force (EMF) of a battery; (ii) the above effect is only revealed when the flux of the magnetic field through the loop changes.[9] Emil Lenz (1804–1865) further discovered[10] that (iii) the magnetic field of the induced current resists the change of the original magnetic field in the loop. Lenz's law indicates that the original magnetic field impedes the motion of the conductive loop that carries the induced current. In other words, the motion of the loop wires stops and the induction of current simultaneously

8. Kittel, *Introduction*, 27–38.

9. Williams, *Michael Faraday*, 151–83.

10. Magie, *Source*, 511–3.

ceases if no energy is continuously put in to drive the motion of the loop wires. Aesthetically, magnetism and electricity are reciprocal: electric field is a driving force (Yang), and magnetic field is an impeding mechanism (Yin).

Faraday-Lenz's law is the last principal discovery which closed up the UCP for a unified theory of electromagnetism, i.e., "electric field by Coulomb's law (YANG) → electric nature of magnetism by Ampere's law (yin) → magnetic field by Gauss's law (YIN) → magnetic field induction of electricity by Faraday-Lenz Law (yang) → electric field by Coulomb's law (YANG)". This UCP will be developed into Maxwell's equations.

Faraday's discovery pushed the application of electric power to the culmination. Battery as the first practical device for electricity is highly limited in energy capacity and power output by the electrochemical mechanism, which makes them too weak to compete with heat engines. But according to Faraday-Lenz's discovery, a much more powerful electromotive force (EMF) can be generated via electromagnetic induction because the induced EMF depends directly on the input power, like heat engines.

The above story is finally concluded in a three-section theory of electrodynamics (Fig. 11), to be elucidated in the following sections.

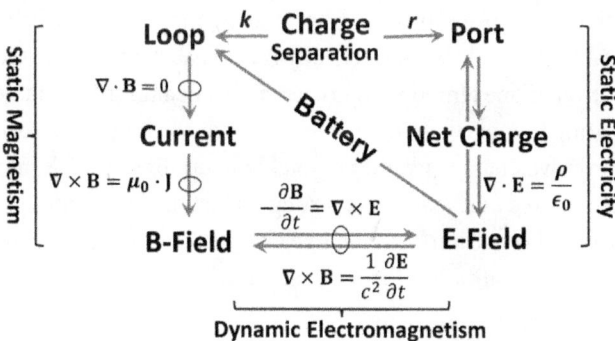

Figure 11. Theoretical structure of electrodynamics. "Loop" is an electric circuit, and "Port" is a capacitor.

Electric Charge: Polar Coding of Matter

We have emphasized that the concealing of charges through charge neutrality forged the suspense in the story of electrodynamics. Because electric charge shows its influence through electric fields, to conceal the electric charges is to conceal electric fields. Then the question is: how can the so strong, diffusive, and long-range electric field be concealed so well in any situations? This question points to the deepest tension between charge polarity and neutrality, which is a long story far beyond the simple arithmetic balance of two types of charges.

In the early 20th century, it became clear that electric polarity was hidden in the chemical building block of matter, that is, the atom. An atom is formed of a massive yet nearly spaceless nucleus in the center and highly light-weighted electrons whose motion is spreading all over the space inside the atom. The nucleus is formed of neutrons with no electric charge and protons each carrying a positive unit charge. Because each electron is carrying a negative unit charge and the number of electrons in an atom is equal to that of protons inside the nucleus, positive and negative charges are balanced. The nucleus holds the identity of the atom, hence, cannot be changed chemically. But the electrons are active and can be exchanged between atoms in chemical reactions driven by the difference of electron affinity of elements. Because of charge neutrality at atomic level, electricity is concealed on a macroscopic scale. Even if one temporarily creates a certain amount of charge by separating the two types of charges in space, the separated positive and negative charges would be quickly neutralized (i.e., homogeneously mixed) driven by the strong attraction between opposite charges and repulsion between like charges. Lightning is a process of charge neutralization through intensive electric current.

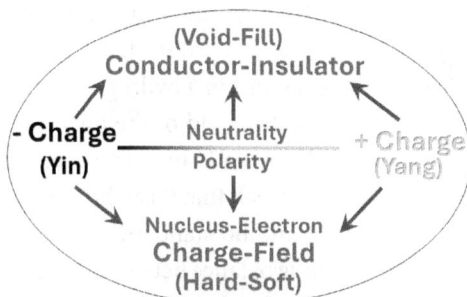

Figure 12. Aesthetic perspective of the polar coding of matter.

The above discovery convinced us that electric charge is the second fundamental quality of matter after mass. Mass is a neutral quality, but charge has polarity. The positive charge has Yang character and negative charge is of Yin. The electric charge as a fundamental quality of matter represents a polar coding of matter (Fig. 12). Regardless of the overall balance of the two types of charges, the Yin-Yang coding of charges governed by the electromagnetic force field forges a tension which polarizes all chemical substance with different capability to hold electrons, as measured by electronegativity[11] and electron affinity.[12] Such polarization plays a crucial role for the physical world, like that of the fundamental tension for all reality. *I Ching* says, "the divine ruler comes forth in Zhen",[13] where "Zhen" is the symbolic image of thunder or lightning. If we take the meaning of Zhen as the tension of polar coding of electric charges, it indeed stimulates many natural processes such as star formation,[14] climate and weather development, chemical reactions, nuclear reactions, etc. The Yin-Yang polarity of electric charges with the electric force field provide the source of energy and basic order of governance for all electric system.

Although electric polarization fights dramatically against charge neutrality on a macroscopic scale to show its power in

11. Pauling, "Nature," 3570–82.
12. Mulliken, "Electroaffinity," 782–93.
13. Lynn, *Changes*, 121.
14. Steinpilz et al., "Electrical," 225–9.

many natural phenomena like lightning, charge neutrality can still beat the charge polarization to conceal electric field at the molecular and atomic level. For example, in a hydrogen molecule (H2) or a helium (He) atom, the electric field of the nuclei are concealed so well that the residual electrostatic interaction between two He atoms or H2 molecules is so weak that it can be overcome, respectively, at 4.2 K and 20 K at around standard pressure. To uncover the mystery behind the subtle balance between charge polarity and neutrality, we must dig more deeply into the basic form of matter.

Notice that charge and field observe the principle of Hard (charge)-Soft (field), in which the positive charge (Yang) can be viewed as the source of field and the negative charge (Yin) the drain. Marvelously, the formulated electric field of point charge preconditions the property of the nucleus and electrons. The massive nucleus is literally a geometric point pinning the center of the atom (Hard), whereas the drastic motion of the highly light electrons is described by spherical harmonic waves (Soft), which as a field of motion, is perfectly adapted to the spreading (Fill) electric field centering the nucleus. This means, the electric field of the nucleus can be drained or screened out (Void) neatly by the mist of electrons to conceal charge polarity perfectly in atomic level.

Compared to all other historically proposed models for atoms in classical picture, such a creation builds an epic tension inside the atoms in perfect harmony, observing all the aesthetic principles. For example, in contrast to the famous plum pudding model[15] that had negatively charged electrons embedded within a positively charged "pudding", the real structure of an atom builds tension in two major "plots". First, the density of positive charge inside the nucleus is 10^{15} times the average negative charge density in the atom, implying an extremely strong binding force must exist to hold the intactness of the nucleus. Second, wave motion of electrons (Soft) confined by the strong electric field results in a whole set of quantum states (Hard) of electrons that encode all chemical elements in the periodic table. These quantum states prevent

15. Thomson, "Structure," 237–65.

electrons from collapsing into the nucleus and are to tell the whole story of the chemical world.

The quantum coding of chemical elements further dictates forms and properties of chemical substances, which are greatly diversified through the interplay of charge polarity and charge neutrality. Here we focus on conductivity and electron affinity, the material properties directly relevant to electrodynamics. Both properties are based on the electronic band structure of materials. When the atoms form solid-state materials, the quantum states of individual atoms will be expanded into energy bands[16] through reciprocal replication in k space: the occupied lower-energy bands of *bonding* states and the higher-energy unoccupied bands of *antibonding* states. The highest occupied energy band (valence band), the lowest unoccupied band (conduction band), and the gap in-between determine the property of a solid. When the two bands overlap, the gap disappears and the solid is a conductor. When the gap is wide (e.g., > 2eV), the solid is an insulator. If the band gap is smaller (e.g. ~ 1.0 eV), the solid is an ideal semiconductor. The property of conductivity observes the principle of Void-Fill. Only the partially occupied (Void) band has conductivity. A fully occupied (Fill) electric band has no conducting electrons because there is no room for charge separation in k space.

Electron affinity of a solid-state material is the energy gained when an electron is transferred to the material from free space or vacuum. The electron affinity of an insulator is normally very small because the conduction band of an insulator is high. That is why it is empty. But surfaces or defects introduce energy states[17] that are lower enough to hold electrons, depending on the chemical property of the material. Because of the open conduction band, conductors can accept and donate electrons. Therefore, in addition to conduction of electric current, conductors can also be used as ports of both positive and negative charges. In order to prevent loss of electric charges or electric power, insulators must be used to separate the conductors from each other or from the environment.

16. Kittel, *Introduction*, 161–82.
17. Kittel, *Introduction*, 493–7.

Static Electricity: from Charges to Fields

The suspenseful story of electromagnetism testifies the paradoxical reality: what we had believed to be more real (e.g., electric materials, charge, and current) was gradually superseded by what we had believed to be virtual (force, field, and potential). As this process goes deeper layer by layer, scientists finally realized that the enigmatic field theory (virtual) turned out to be the most beautiful quest for the truth hinted initially by the charge particles (real).

Point Charges and Coulomb Force

An electrostatic generator based on triboelectric effect has a central component such as a rotating glass ball or disk with its surface in close contact with fixed woolen or rubber pad. The friction between the two materials electrifies them with opposite charges. These electric materials are insulators rather than conductors, because electric charges created in conductors would be quickly conducted away and dissipated to the "ground" for neutralization. In contrast, the charges accumulated on the surface of an insulator can build up a high electric potential because the conduction band of the insulator is too high to hold many electrons. Conductors then found their application to conduct the generated charges to a charge-storage device called Leyden Jar,[18] an early form of capacitors. With stable storage of electric charges and the elaborate use of conductors, systematic scientific experimentation can be conducted for in-depth study of the nature of electricity. Conductors started to play the central role, being used as ports or carrier of electric charges for charge conduction, division, and distribution. The study of positive and negative charges, attraction and repulsion between the two types of charges, and electrostatic induction, etc. led to the discovery of Coulomb's law,[19] which touched the foundation of electromagnetism.

18. Heilbron, *Electricity*, 309–21.
19. Heilbron, *Electricity*, 449–73.

Coulomb's law describes the interaction between two point-charges separated by a certain distance. Like its counterpart of the mass point in mechanics, point charge is ideal model of a charge holder such as a metal ball whose radius is much smaller than the considered distance of interaction between such balls. With this basic model, the force model is given in the mathematical form of Coulomb's law. Therefore, the interaction between charges is called Coulomb force, or more generally, electromagnetic force. Just like the universal gravitation is based on mass, Coulomb force is based on electric charge. The Coulomb force also observes the inverse-square law, $F = kq_1q_2/r^2$, where $k = 8.988 \times 10^9$ N·m^2/C^2 is the force constant, q_1 and q_2 are the charge quantities of the two point-charges, r is the distance between them. Unlike the gravitational force that always makes two mass particles attract each other, the Coulomb force between two charge particles can be either attractive or repulsive: like charges repel and opposite charges attract. Both universal gravitation and Coulomb force are fundamental interactions. Although mass is the most tangible quantity in macroscopic scale, the action of gravitational force between normal earthly bodies (excluding the Earth itself) is hidden. Electric charge is normally hidden in macroscopic scale, but Coulomb force dictates both the structure of atoms and the macroscopic forms of bulk materials. Coulomb force is much stronger than gravitational force. If two men are 1 meter apart, each holding 10 mC (10^{-3} C) electric charges, the Coulomb force between them would be nearly 2000 lb (9000 N). If each man measures 60 kg, the gravitational force between them is only 10^{-6} lb.

The similarity of Coulomb force with universal gravitation is a reminiscence of the "action at a distance". Therefore, the discovery of Coulomb force further appeals for deeper understanding of non-contact conservative forces. Field theory developed for gravitation in post-Newtonian era was largely mathematic (virtual). It is through the study of electromagnetism that fields became known as a fundamental reality. The invention of Faraday shield[20] based on electrostatic induction clearly shows that an electric field can be

20. Giancoli, *Physics*, 457.

screened out of a particular spatial region. Convincingly, the electric force is directly due to the field rather than the electric charge.

Electric Fields: Refinement of "Action at a Distance"

Aesthetically, charge particles are real objects, and a field is a virtual object. Therefore, the study of force fields should be in a nontangible (virtual) way which relies on rigorous mathematic description of fields using vector analysis. Remarkably, Faraday invented force lines[21] as a tangible (real) representation for invisible fields. A brief comparison of the two types of methods allows us to taste the aesthetic flavor of electricity. Although the two methods appear to be distinctively different, they compensate each other so beautifully: the field lines depict a transparent picture for intuitive understanding, whereas the field theory developed by James Clerk Maxwell[22] turns Faraday's masterpiece of scientific artwork into meticulous rigorous mathematics. In fact, all the key components of Maxwell's theory can be traced back to the ideas in Faraday's works.[23]

The electric field related to motionless charges is static electric field. The elegant mathematic expression of Coulomb's law hints the electric field theory: the force scales linearly with the two charges, which is mathematically the prerequisite of (i) separation of field from probing charge, and (ii) supposition of electric fields from multiple source charges; and the rigorous inverse square law of the central force ensures (iii) the conservation of field flux of the given charge source.

The electric field at a particular point of space, E, is a vector with both its magnitude and direction depending on the location in space. Although the electric field is not visible, it can be probed or measured by a point charge with tiny amount of charge, q, called a probing charge. This is because the force of an electric field exerted

21. Hesse, *Forces and Fields.* 198–205.

22. Maxwell, "Physical Lines," 11–23.

23. Joseph, *Lightning*, 104–5.

to the small probing charge is always proportional to the charge $F = qE$. Such a property of linear scaling is rooted in Coulomb's law. If Coulomb's law had a different formula, e.g., $F = k(q_1+q_2)^2/r^2$, the definition of $F = qE$ would not be feasible. According to the true Coulomb's law, the stronger the electric field, the bigger the force exerted by the field to a certain probing charge. Therefore, the magnitude of the electric field is measured by $E = F/q$. The density of the field lines at a location represents the magnitude of the field. The direction of the field is marked by an arrow along the tangent to the force line, pointing from positive to negative charge.

According to spherical symmetry of Coulomb's law, the electric field from a source charge Q is a central field pointing along the radial direction, with the magnitude observing the inverse square law. In Faraday's representation of field lines (Fig. 13), all the field lines originate from the source charge Q and the total number of field lines conserves no matter how far they run away from the source charge. This is the result of inverse square law and why the charge is called the source of the field.

In field theory, such a feature is described meticulously in the differential equation that relates the field vector E with the source charge Q, as shown in Fig. 13. The divergence of field is the differential flux per unit volume through the closed surface that encloses the differential volume with the charge density ρ.

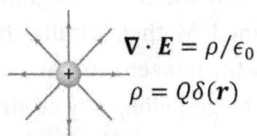

$$\nabla \cdot E = \rho/\epsilon_0$$
$$\rho = Q\delta(r)$$

Figure 13. Electric field of a point charge Q represented in illustration of field lines (Real) and the differential equation or Gauss law in field theory (Virtual).

Energy Landscape of an Electric Field

A more convenient way to map a field is the landscape of potential or potential energy which is a scalar. This is only possible when the field is a conservative force.

As we have discussed in chapter 3, the work of a conservative force does not depend on the path of motion. It only depends on the initial and final position, which allows for definition of scalar function of space called potential energy. The work done by the conservative force is then potential energy difference between the two positions. Indeed, static electric fields are conservative. The potential energy of a probing charge in the field depicts the energy landscape of the charge in the electric field. The electric potential energy of a charge q in the electric field is proportional to the charge, qV, where V is called electric potential. Potential is a scalar measured in the unit of volt (V) or joules per coulomb (J/C). Like field strength E, the potential V is the property of the field, independent of probing charge q.

Because an electric field has a source charge, its force lines originate from positive charge and end at negative charge. Generally, such a field cannot simultaneously have curl structure with closed loop. As is shown in Fig. 14, the line marked with a cross cannot be a real field line. This is because a field line cannot have an overlap or cross point where the magnitude and/or direction of the field is not defined. Mathematically, the loop structure of a field line is defined by the nonzero curl of the field. For a field that has no curl, its path integral along any arbitrary closed loop, e.g., AaBbA in Fig. 14, is always zero. That is, the integral from A to B only depends on locations A and B, disregarding the path. Such a path integral defines a potential for the field, which depicts the potential energy landscape of any arbitrary probing charge in the field.

$$\nabla \cdot E = \rho/\epsilon_0 \qquad \oint_{Aa}^{B} E \cdot dl = \oint_{Ab}^{B} E \cdot dl$$

$$\nabla \times E = 0 \longrightarrow E = -\nabla\varphi$$

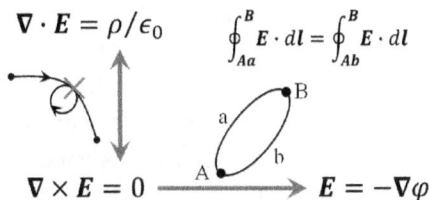

Figure 14. The property of conservative force field. It has no curl and can be defined by the divergence of a scalar potential function because the work done by the force does not depend on pathways.

Towards Useable Electric Energy

The electric field as a conservative force field can store energy. For example, in lightning formation, the kinetic energy or heat energy of the clouds is converted into electric potential energy stored in the thunderclouds. Charge separation in the thundercloud stores a huge amount of electric energy and builds up a high potential difference. When the potential difference is big enough, a conductive channel of plasma between the two oppositely charged regions can be formed for charge neutralization. The electric potential energy released could reach the order of 10^9 J in less than a second,[24] transforming into heat, energy of sound waves, and light emission etc. Such extremely high power mobilizes the circulation of materials in the natural environment for renewal of natural resources.[25]

The charged thunderclouds form an effective capacitor in gigantic spatial and energy scale. All capacitors use charge separation to store electric potential energy directly. Parallel plates capacitors are the simplest model of a capacitor, in which two oppositely charged flat metal plates are separated by a layer of insulating material called dielectric. Normally the area of plates, A, is much bigger than the separation, d, between the plates. In this situation, the electric field pointing from the positive plate to

24. Giancoli, *Physics*, 504.
25. Brune et al, "Extreme", 711–5.

the negative one is considered to have the same magnitude and direction everywhere. The electric field outside the capacitor can be neglected due to screening effect. Capacitance (in farad, F) is the intrinsic property of a capacitor and is defined by the stored charge Q per volt of potential difference V. A parallel plate capacitor separated by air has the capacitance $C = \epsilon_0 A/d$. Based on field theory, the energy density of the electric field inside the capacitor is $\epsilon_0 E^2/2$.

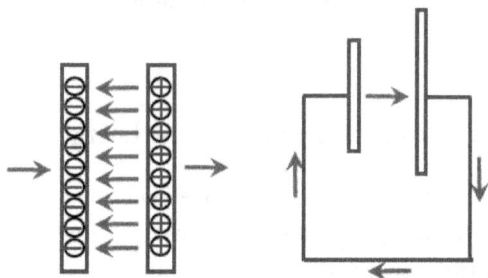

Figure 15. Comparison of a capacitor (left) with a battery (right). In contrast to the electric field starting positive charges and ending on negative charges, the electric field generated by a battery is closed loop along the electric circuit.

Like lightning, the electric energy stored in capacitors cannot be released steadily. In contrast, batteries store chemical energy and release it steadily into electric energy to power a load. Unlike the energy storage in capacitors through charge separation, the energy stored in batteries is chemical energy of the reactants relative to that of the products of an electrochemical reaction. Chemical energy provides electromotive force (EMF) that drives charge flow in a closed conductive circuit without breaking charge neutrality. The invention of batteries was theoretically radical because the EMF of a battery creates a looped electric field with no source charge hence with a nonzero curl, as shown in Fig. 15, in sharp contrast to the static electric field caused by charge separation in a capacitor.

If a light bulb is connected into the loop through conducting wires, electric current in the wires will be driven to flow across the light bulb and light it up. According to Ohm's Law, the output potential difference V of the battery is proportional to the current, I, passing through the light bulb, $V = I R$, with the scaling constant R being the resistance of the light bulb. The power output is $P = V I$.

Static Magnetism

Magnetism (Yin) is the reciprocal replication of electricity (Yang). This aesthetic setting is manifested splendidly in God's creation: electricity is revealed from the sky (Yang) in lightning, but magnetism is detected in the global geomagnetic field emanating from the depth of the earth (Yin). Electricity is realized through charge separation in real space, yet magnetism is induced by charge separation in k space.

Magnetism: The Field without a Source

Unlike electric charges that were generated through triboelectric effect, magnets were found to exist in natural forms such as lodestone[26] which are alloys of certain transition-metals like iron, cobalt, and nickel. The south pole of a suspended needle-like magnet points to the south due to its interaction with the geomagnetic field emanating from the Earth.

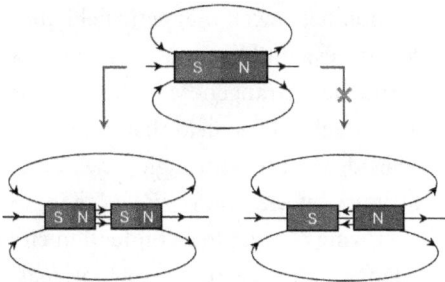

Figure 16. Magnetic field lines enter inside the magnet forming closed loops. It is not possible to obtain monopoles by dividing a bar magnet.

26. Benjamin, *History of Electricity*, 15–20.

A magnet has two opposite poles called the north pole and south pole. Like poles repel and opposite poles attract through magnetic field. Magnetic field B can also be depicted with field lines. Unlike charges, one can never separate the two poles of a magnets, no matter what tiny pieces a magnet is divided into. This disapproves any hypothetical sources of B fields, so-called monopoles. Figure 16 shows the magnetic field lines of the bar magnet (top). When it is cut into two smaller magnets, the magnetic field lines in between run the same way as the external field lines (bottom). If otherwise the magnet is cut into two monopoles, these field lines should point against the external lines. This observation convinces us that a magnetic field line always forms a closed loop. To translate this Faraday's picture into field theory, the divergence of the magnetic field should be zero. Because of this, a magnetic field must have nonzero curl, which reveals the electric origin of magnetism.

Electric Origin of Magnetism

The similarity between electricity and magnetism suggests potential unification of the two based on aesthetic replication. But the strangeness of the loop structure of magnetic field showing the lack of source imposed a seemingly insurmountable challenge to the task for a long historical period. Such strangeness had blocked all clues of interaction between a magnetic field and a static electric field or static charges, until the new strangeness in conformity was discovered. This new strangeness is the looped electric current driven by a circuital electric field that is compatible with the curling magnetic field. The breakthrough benefited crucially from the invention of battery by Alessandro Volta (1745–1827),[27] which generate steadily moving charges in a conduction circuit.

When constant electric current in a loop was created, the mysterious relationship between electricity and magnetism was quickly unveiled. In 1820, Oersted found that electric current

27. Joseph, *Lightning*, 35–40.

induces a magnetic field.[28] This is to say, magnetism is not a fundamental quality of matter; instead, it is the derived property of electric current. Soon after that, Andre-Marie Ampere (1775–1836) proposed a model[29] which assumes the elementary unit of magnet is a molecular current. We now understand that the so-called molecular current is due mainly to the orbital and spin of electrons inside the atoms. Oersted's discovery inspires a new wave of study on the electric nature of magnetism. Jean-Baptiste Biot and Felix Savart developed the Biot-Savart theorem[30] of magnetic field B induced by electric current. Ampere developed a general formula to determine the force between flows of electric current, which is the magnetic counterpart of Coulomb's law. These series of efforts are concluded by the differential form of Ampere's circuital law.

A simpler form of Ampere's law is that the magnetic field B produced at a particular location by a constant current I in a long conductive wire is inversely proportional to the distance (r) from the wire: $B = \mu_0 I/(2\pi r)$. The constant μ_0 is called the permeability of free space, and it has the value: $\mu_0 = 4\pi \times 10^{-7}$ T·m/A.

Dynamics of Electromagnetism

Inspired by Oersted's work, Faraday started a whole series of experiments on the production of electricity from magnetism.[31] His initial failure to generate electric current using a static magnetic field was like an attempt using a magnet to act on static charge. Surprisingly yet most fortunate, a changing magnetic field induces a curling electric field which is exactly an EMF that powers an electrical load.

28. Oersted, "Experiments," 273–6.

29. Williams, *Michael Faraday*, 143–50.

30. Jackson, *Classical Electrodynamics*, 169–72.

31. Williams, *Michael Faraday*, 143–50.

Faraday's Law of Electromagnetic Induction

In the field-line representation, Faraday's law of electromagnetic induction states that a changing magnetic field induces an electric field whose field lines form a closed loop, and the induced electric potential equals the changing rate of the flux of the magnetic field passing through the loop. The induced electric potential causes electric current in the wire loop; and, according to Ampere's law, this induced current also induces a magnetic field superposing on the original field. Lenz's discovery characterizes the Yin character of the magnetic field and its impeding of change.

According to Faraday-Lenz law the faster a magnetic field is forced to change, the larger the induced EMF, the larger would be the corresponding current in a conduction circuit, and the larger the induced impeding magnetic field. Such negative feedback requires that the output power of an electric generator be determined by the input power that forces the magnetic field to change. Faraday's work laid the technological foundation for electric power generators of alternating electric current. It is also crucial for long-distance delivery of electric power through transformers.

However, a close examination of how Faraday-Lenz Law is coordinated with Ampere's Law exposes a hidden issue in the latter. In the above reasoning, we assume that the impedance of a magnetic field to its changing by the induced EMF must be realized by the EMF-driven current, which further induces a negative magnetic field according to Ampere's law. But what if there is no circuit loop to conduct an electric current along the induce EMF? If the impedance must be there without the induced current, the electric field of the induced EMF should also be able to induce a negative magnetic field. This implies an electric field is missing in Ampere's law. This issue is seen more clearly from the aesthetic perspective that the laws of Gauss, Ampere, Faraday-Lenz observe the UCP of "Gauss (Coulomb's) law for electricity → Ampere's circuital law (electricity to magnetism) → Gauss law for magnetism → Faraday-Lenz law (magnetism to electricity) → Gauss law for electricity". The electric field is YANG, magnetic field is YIN. Then Ampere's law is *yin* and Faraday-Lenz law is *yang*. However, strangeness can be recognized in the apparently perfect harmony of the UCP, because the electric field induced by the changing magnetic field is a looped one. That means the Gauss law of electricity in the UCP must be written in the same form with no divergence as that for magnetism, whereas the Gauss law for static electricity should be treated as an addition outside of the main UCP loop. Similarly, the original Ampere's law without involving an electric field must also be displaced as an additional term superposed on a missing equation describing the induction of magnetic field by the electric field, as shown in Fig. 17. It turns out that the missing equation was soon added by Maxwell in his unified theory of electrodynamics.

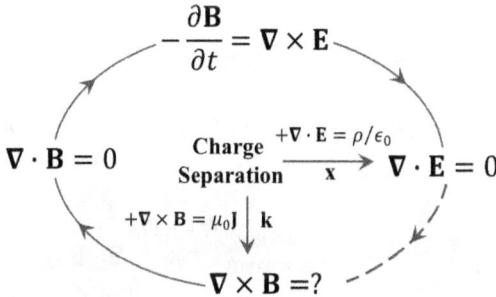

Figure 17. The UCP of the fundamental laws of electrodynamics. Static electricity (Coulomb's law) and static magnetism (Ampere's law) based on charge separation in x and k spaces are displaced from the main loop of the UCP as strangeness. The unknow equation should be added to show how a magnetic field is induced by an electric field in free space without electric current.

Maxwell's Theory

What Maxwell did was to add an electric field term to Ampere's law (see Fig. 18), called generalized Ampere's law or Ampere-Maxwell law, which was installed as a fundamental hypothesis for electrodynamics. Notice that the original Ampere's law is only valid for static magnetism and steady current. The Ampere-Maxwell law is applicable to dynamic magnetism and changing current. The four equations in Fig. 18, called Maxwell's equations because they are broader than their original meaning for static electromagnetism, are connected seamlessly in a complete UCP. Maxwell's equations are dynamics equations, serving as four fundamental laws of electrodynamics. Figure 18 also shows that the UCP of the four Maxwell's equations is more neatly symmetric without net charge and current. Because the changing electric field can induce a negative magnetic field fed back to the original magnetic field through Faraday-Lenz law, a sinusoidal oscillation of field will carry the cyclic induction on and on so that energy can be spread out in free space. This is the prediction of electromagnetic waves. Maxwell's theory further predicts that light is an electromagnetic wave based on the universal constant $c = (\epsilon_0 \mu_0)^{-1/2}$ as the speed of both light and electromagnetic waves in free space.

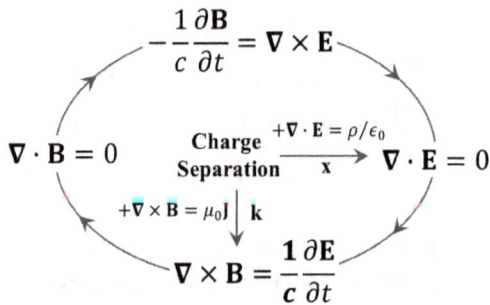

$$-\frac{1}{c}\frac{\partial \mathbf{B}}{\partial t} = \mathbf{\nabla} \times \mathbf{E}$$

$$\mathbf{\nabla} \cdot \mathbf{B} = 0$$

Charge Separation $+\mathbf{\nabla} \cdot \mathbf{E} = \rho/\epsilon_0$ $\mathbf{\nabla} \cdot \mathbf{E} = 0$

$+\mathbf{\nabla} \times \mathbf{B} = \mu_0 \mathbf{J} \mid \mathbf{k}$

$$\mathbf{\nabla} \times \mathbf{B} = \frac{1}{c}\frac{\partial \mathbf{E}}{\partial t}$$

Figure 18. The maxwell equations in UCP.

This discovery represents the second most important synthesis of physics after Newton's grand synthesis in mechanics. However, it is in this prediction that a tension is exposed between Maxwell equations and the Newton-Galileo principle of relativity: if a speed can be a universal constant, it defines absolute motion. It is this tension that pushes physics towards a revolution. To reconcile the contradiction, Einstein proposed that the speed of light in free space must be treated as a supreme constant so that the principle of relativity still holds for motion of all other objects. This is the basic postulate of the special theory of relativity.[32]

32. Einstein, *Relativity*, 14–36.

6

Summary and Prospects

A FORMULATED APPROACH TO aesthetic criticism of science is extracted from *I Ching*. Its informing power for modern science is demonstrated in criticism of the three theories of classical physics i.e., mechanics, thermodynamics, and electrodynamics. Finally, we should think about how these three theories provide a whole picture of the macroscopic world following the right aesthetic order. To this end, the three theories should be overviewed as a whole system through the five aesthetic axioms.

Physics and modern science develop along the axis of paradoxical reality: matter was regarded as real, and action or force was considered virtual; however, as it went deeper, fundamental force fields as the origin of action were found to be the primary reality which not only sustains the forms of matter but also creates matter. This is reflected in the trajectory of classical physics from Newtonian mechanics to Maxwell theory, and their transition from classical to modern physics. In Newtonian mechanics, a mass particle as the only object of central concern was an absolute physical entity (Real) in contrast to the conceptual universal gravitation (Virtual) being treated as an action at a distance. The ideal model of mass point allows for Newtonian mechanics to inherit the elegance and transparency of Euclidean geometry by circumventing dealing with any complex forms and internal structures of matter. Continuity and homogeneity describe the basic attribute of both matter and space. It is truly marvelous that these simplified concepts sketch a closed self-consistent theory of fundamental physics, which is featured by the deterministic doctrine of motion under the aesthetic axioms of Hard-Soft. In contrast, thermodynamics is featured by Void-Fill: an assembly of tremendous number of particles fill the phase space, on which the changeability of a thermodynamics system and all the central concepts (i.e., temperature, entropy, and heat) are based. Because it is impossible to deal with individual particles, the concept of mass particle is now "virtualized" as mathematical objects; and the thermodynamic potentials due to the filling status of the phase space play the decisive role of the driving force. The theory of electrodynamics is featured by Yin-Yang: not only matter particles are polarized into positive and negative charges but the force decomposed into electric (Yang) and magnetic (Yin) fields. This theory proves for the first time that force fields are totally real and can be handled experimentally. Modern quantum field theory further shows that even mass is created by field.

This study sets the basic framework of formulated aesthetics for science and art. As an advanced criticism approach, it inspires critical thinking for advancing fundamental research. Compared with the systematic philosophical criticism[1] that is formidable for most scientists, aesthetic approach is intuitively more accessible. Tethered to the origin of beauty and the way beauty is expressed in the viewing angle of Tao, the principles installed here are expected to promote organic creativity and objective judgement. Therefore, future study should also be directed towards the development of algorithms and parameterized models, which can equip artificial intelligence (AI) for more authentic artistic creation and creative scientific exploration.

1. See Popper, *Conjectures and Refutations*; and Kuhn, *The Structure of Scientific Revolutions*.

Bibliography

Barnes, Jonathan. *The Presocratic Philosophers*. New York: Routledge Taylor & Francis, 1982.

Benjamin, Park. *A History of Electricity (the Intellectual Rise in Electricity): From Antiquity to the Days of Benjamin Franklin*. New York: John Wiley & Sons, 1898.

Bernoulli, Daniel. "Hydrodynamica (1738)," in *Landmark Writings in Western Mathematics 1640-1940*. Translated by G. K. Mikhailov, edited by I. Grattan-Guinness, 131-42. Elsevier, 2005.

Breitenbach, Angela. "Aesthetics in Science: A Kantian Proposal," *Proceedings of the Aristotelian Society*, New Series, 113 (2013) 83-100.

Brett, Christopher M. A. and Brett, Ana M. O. *Electrochemistry: Principles, Methods, and Applications*. Oxford: Oxford University Press, 1993.

Brune, McFarland, et al. "Extreme Oxidant Amounts Produced by Lightning in Storm Clouds." *Science* 372 (2021) 711-15.

Cajori, Florian. "History of Zeno's Arguments on Motion: Phases in the Development of the Theory of Limits," *The Am. Mathematical Monthly* 22 (1915) 292-97.

Chandrasekhar, S. *Truth and Beauty: Aesthetics and Motivation in Science*. Chicago: The University of Chicago Press, 1987.

Chin, Annping. *The Analects of Confucius*, Book 9.17. Translated by Annping Chin. New York: Penguin Books, 2014.

Clausius, Rudolf. *Mechanical Theory of Heat with Its Application to the Steam Engine and to the Physical Properties of Bodies*. Edited by T. Archer Hirst, F. R. S., London: John Van Voorst, 1867.

Cooper, Leon N. *Physics: Structure and Meaning*, Hanover: University Press of New England, 1992.

Dexter, A. R. "Advances in Characterization of Soil Structure," *Soil and Tillage Research* 11 (1988) 199-238.

Dirac, Paul A. M. "The Evolution of the Physicist's Picture of Nature," *Scientific American* 208 (1963) 45-53.

———. "The Excellence of Einstein's Theory of Gravitation," in *Einstein: The First Hundred Years*, edited by Maurice Goldsmith, Alan Mackay, and James Woudhuysen, 41–46. Oxford: Pergamon, 1980.

Einstein, Albert. *Investigations on the Theory of Brownian Movement.* New York: Dover Publications, 1956.

———. *Relativity: The Special and General Theory.* Translated by Robert W. Lawson. New York: H. Holt, 1920.

Elgin, Catherine Z. "Fiction as Thought Experiment," *Perspectives on Science* 22 (2014) 221–41.

Engler, Gideon. "Aesthetics in Science and in Art," *British Journal of Aesthetics* 30 (1990) 24–34.

Field, G. C. "Plato and Natural Science," *Philosophy* 8 (1933) 30–41.

Friedman, Michael. *Reconsidering Logical Positivism.* Cambridge: Cambridge University Press, 1999.

Gendler, Tamar S. "Thought Experiments Rethought—and Rreperceived," *Philos. of Science* 71 (2004) 1152–63.

Giancoli, Douglas C. *Physics: Principles with Applications*, 6th Edition, Upper Saddle River: Pearson, 2005.

Godfrey-Smith, Peter. *Theory and Reality: An Introduction to the Philosophy of Science.* Chicago: University of Chicago Press, 2010.

Hardie R. P., and Gaye R. K. *Physics Book VI, by Aristotle, Written 350 B.C.E*, The Internet Classics Archives https://classics.mit.edu/Aristotle/physics.6.vi.html.

Heilbron, J. L. *Electricity in the 17th and 18th Centuries: A Study of Early Modern Physics.* Mineola, NY: Dover, 1979.

Heisenberg, Werner. *Across the Frontiers.* Translated by Peter Heath. New York: Harper and Rows, 1974.

Herapath, John. "On the Physical Properties of Gases," *Annals of Philosophy* 8 (1816) 56–60. https://books.google.com/books?id=dBkAAAAAMAAJ&pg=PA56.

Hesse, Mary B. *Forces and Fields: The Concept of Action at a Distance in the History of Physics.* West Port: Greenwood, 1962.

Jackson, John D. *Classical electrodynamics.* 2nd edition. New York: John Wiley & Sons, 1975.

Jaspers, Karl. *The Origin and Goal of History.* Translated by Michael Bullock. London: Routledge Classics, 2021.

Joseph, Kathy. *The Lightning Tamers.* Smart Science, 2022.

Joule, James P. "On the Mechanical Equivalent of Heat," *Philosophical Transactions of the Royal Society of London.* 140 (1850) 60–82. https://archive.org/details/philtrans00608634/page/n21/mode/2up?view=theater.

Kittel, Charles. *Introduction to Solid State Physics.* 8th Edition, New York: John Wiley & Sons, 2005.

Kivy, Peter. "Science and Aesthetic Appreciation," *Midwest Studies in Philosophy* 16 (1991) 180–95.

Kline, Morris. *Mathematical Thought from Ancient to Modern Times*. New York: Oxford University Press, 1972.

Kubas, Gregory J. "Metal–dihydrogen and σ-bond Coordination: the Consummate Extension of the Dewar–Chatt–Duncanson Model for Metal–olefin π Bonding," *J. Organomet. Chem.* 635 (2001) 37–68.

Kuhn, Thomas S. *The Structure of Scientific Revolutions*. 3rd Edition, Chicago: University of Chicago Press, 1996.

Lancaster-Brown, Peter. *Halley and His Comet*. Blanford, 1985.

Leibniz, Gottfried W. *New Essays on Human Understanding*. Translated by Peter Remnant and Jonathan Bennett. Cambridge: Cambridge University Press, 1981.

Liboff, Richard L. "The Correspondence Principle Revisited," *Physics Today* 37 (1984) 50–55.

Lynn, Richard J. *The Classic of Changes, A New Translation of the I Ching as Interpreted by Wang Bi*. Translated by Richard J. Lynn. New York: Columbia University Press, 1994.

Magie, William F. *A Source Book in Physics*. Cambridge: Harvard University Press, 1963.

Mair, Victor H. *Tao Te Ching: The Classic Book of Integrity and the Way*. Translated by Victor H. Mair. New York: Bantam Books, 1990.

Maxwell, James C. "On Physical Lines of Force," *Philosophical Magazine* 90 (1861) 11–23.

McAllister, James W. *Beauty and Revolution in Science*. New York: Cornell University Press, 1996.

———. "Is Beauty a Sign of Truth in Scientific Theories? Why are Some New Theories Embraced as Beautiful, Others Spurned as Ugly? Progress in Science May Require That Aesthetic Ideals Themselves Change," *American Scientist* 86 (1998) 174–83.

McDonald John H. *Tao Te Ching*. Translated by John H. McDonald. London: Arcturus, 2010.

Melo, Ivan. "Aesthetic Criteria in Fundamental Physics—The Viewpoint of Plato," *Philosophies* 7 (2018) 96.

Mizrahi, Moti. "Idealizations and Scientific Understanding," *Philos. Stud.* 160 (2012) 237–52.

Mulliken, Robert S. "A New Electroaffinity Scale; Together with Data on Valence States and on Valence Ionization Potentials and Electron Affinities." *J. Chem. Phys.* 2 (1934) 782–93.

Newton, Isaac. *Sir Isaac Newton's Mathematical Principles of Natural Philosophy and His System of the World*. Translated by Andrew Motte, edited by Florian Cajori. Berkeley: University of California Press, 1962.

———. *Isaac Newton's Papers & Letters on Natural Philosophy and Related Documents*, edited by I. Bernard Cohen and Robert E. Schofield. Cambridge: Harvard University Press, 1958.

Norton, John D. "Nature is the Realization of the Simplest Conceivable Mathematical Ideas: Einstein and the Canon of Mathematical Simplicity," *Stud. Hist. Philos. Mod. Phys.* 31 (2000) 135–70.

Oersted, Hans C. "Experiments on the Effect of a Current of Electricity on the Magnetic Needle," *Annals of Philosophy*, 16 (1820) 273–76.

Pauling, Linus. "The Nature of the Chemical Bond. IV. The Energy of Single Bonds and the Relative Electronegativity of Atoms," *J. Am. Chem. Soc.* 54 (1932) 3570–82.

Penrose, Roger. "The Role of Aesthetics in Pure and Applied Mathematical Research," *Bulletin of the Institute of Mathematics and Its Applications* 10 (1974) 266–71.

Poincaré, Herry. *The Foundations of Science.* Translated by George B. Halsted. New York: Science, 1913.

Polkinghorne, John C. *Serious Talk: Science and Religion in Dialogue.* Valley Forge, PA: Trinity Press International, 1995.

Popper, Karl R. *Conjectures and Refutations.* New York: Basic Books, 1962.

Rota, Gian-Carlo. "The Phenomenology of Mathematical Beauty," *Synthese* 111 (1997) 171–82.

Shih, Vincent Y. *The Literary Mind and the Carving of Dragons by* LIU HSIEH. Translated by Vincent Y. Shih. New York: The Columbia University Press, 1959.

Steinpilz, Tobias and Joeris, Kolja. et al., "Electrical Charging Overcomes the Bouncing Barrier in Planet Formation." *Nature Physics.* 16 (2020) 225–9.

Stuart, Michael T. "Thought Experiments," in *The Routledge Companion to Thought Experiments,* edited by Michael T. Stuart, Yiftach Fehige, and James R. Brown, 526–44. London: Routledge, 2018.

Tate, Alen. *The man of Letters in the Modern World, Selected Essays, 1928–1955.* New York: Meridian Books, 1955.

Thompson, Benjamin. "XV. New Experiments Upon Gun-Powder, With Occasional Observations and Practical Inferences; To Which Are Added, an Account of a New Method of Determining the Velocities of All Kinds of Military Projectiles, and the Description of a Very Accurate Eprouvette for Gun-Powder," *Philosophical Transactions of the Royal Society of London.* 71 (1781) 229–328. https://doi.org/10.1098%2Frstl.1781.0039.

Thomson, J. J. "XXIV. On the Structure of the Atom: An Investigation of the Stability and Periods of Oscillation of a Number of Corpuscles Arranged at Equal Intervals around the Circumference of a Circle; with Application of the Results to the Theory of Atomic Structure," *The London, Edinburgh, and Dublin Philosophical Magazine and Journal of Science* 7 (1904) 237–65.

Todd, Cain. "Unmasking the Truth Beneath the Beauty: Why the Supposed Aesthetic Judgements Made in Science May Not Be Aesthetic at All," *International Studies in the Philosophy of Science* 11 (2008) 61–79.

Walhout, Peter K. "The Beautiful and the Sublime in Natural Science," *Zygon* 44 (2009) 757–76.

Williams, L. Pearce. *Michael Faraday.* New York: Basic Books, 1965.

Wolterstorff, Nicholas. *Art in Action: Toward a Christian Aesthetic*. Grand Rapids: Eerdmans, 1987.

Zangwill, Nick. *The Metaphysics of Beauty*. Ithaca: Cornell University Press, 2001.

Zee, Anthony. *Fearful Symmetry: The Search for Beauty in Modern Physics*. Princeton: Princeton University Press, 1999.

Zhu, Xi. *Commentary of the Doctrine of Mean* in *The Collected Commentaries of the Chapters and Verses of the Four Books*, in Chinese. https://www.zhonghuadiancang.com/rulizhexue/zhongyongzhangju/3351.html.

Ziporyn, Brook. *Zhuangzi: The Essential Writings and Selections from Traditional Commentaries*. Translated by Brook Ziporyn. Indianapolis/Cambridge: Hackett, 2009.

www.ingramcontent.com/pod-product-compliance
Lightning Source LLC
Chambersburg PA
CBHW052149090426
42741CB00010B/2204